PLANNING FOR THE COMMON GOOD

Appeals to the 'common good' or 'public interest' have long been used to justify planning as an activity. While often criticised, such appeals endure in spirit if not in name as practitioners and theorists seek ways to ensure that planning operates as an ethically attuned pursuit. Yet, this leaves us with the unavoidable question as to how an ethically sensitive common good should be understood. In response, this book proposes that the common good should not be conceived as something pre-existing and 'out there' to be identified and applied or something simply produced through the correct configuration of democracy. Instead, it is contended that the common good must be perceived as something 'in here,' which is known by engagement with the complexities of a context through employing the interpretive tools supplied to one by the moral dimensions of the life in which one is inevitably embedded. This book brings into conversation a series of thinkers not normally mobilised in planning theory, including Paul Ricoeur, Alasdair MacIntyre and Charles Taylor. These shine light on how the values carried by the planner are shaped through both their relationships with others and their relationship with the 'tradition of planning' – a tradition it is argued that extends as a form of reflective deliberation across time and space. It is contended that the mutually constitutive relationship that gives planning its raison d'être and the common good its meaning are conceived through a narrative understanding extending through time that contours the moral subject of planning as it simultaneously profiles the ethical orientation of the discipline. This book provides a new perspective on how we can come to better understand what planning entails and how this dialectically relates to the concept of the common good. In both its aim and approach, this book provides an original contribution to planning theory that reconceives why it is we do what we do, and how we envisage what should be done differently. It will be of interest to scholars, students and practitioners in planning, urban studies, sociology and geography.

Mick Lennon is Associate Professor of Planning and Environmental Policy in the School of Architecture, Planning and Environmental Policy at University College Dublin, Ireland.

THE RTPI Library Series
Editors: Akira Drake Rodriguez
Weitzman School of Design, University of Pennsylvania
Mark Tewdwr-Jones
Bartlett Centre for Advanced Spatial Analysis, University College London
Belinda Yuen
Lee Kuan Yew Centre for Innovative Cities, Singapore University of Technology and Design

Published by Routledge in conjunction with The Royal Town Planning Institute, this series of leading-edge texts looks at all aspects of spatial planning theory and practice from a comparative and international perspective.

Public Norms and Aspirations
The Turn to Institutions in Action
Willem Salet

Planning, Law and Economics
The Rules We Make for Using Land, second edition
Barrie Needham, Edwin Buitelaar and Thomas Hartmann

A Future for Planning
Taking Responsibility for Twenty-First Century Challenges
Michael Harris

Digital Participatory Planning
Citizen Engagement, Democracy, and Design
Alexander Wilson and Mark Tewdwr-Jones

Planning for the Common Good
Mick Lennon

For more information about this series, please visit: www.routledge.com

PLANNING FOR THE COMMON GOOD

Mick Lennon

NEW YORK AND LONDON

First published 2022
by Routledge
605 Third Avenue, New York, NY 10158

and by Routledge
2 Park Square, Milton Park, Abingdon, Oxon, OX14 4RN

Routledge is an imprint of the Taylor & Francis Group, an informa business

© 2022 Mick Lennon

The right of Mick Lennon to be identified as author of this work has been asserted by him in accordance with sections 77 and 78 of the Copyright, Designs and Patents Act 1988.

All rights reserved. No part of this book may be reprinted or reproduced or utilised in any form or by any electronic, mechanical, or other means, now known or hereafter invented, including photocopying and recording, or in any information storage or retrieval system, without permission in writing from the publishers.

Trademark notice: Product or corporate names may be trademarks or registered trademarks, and are used only for identification and explanation without intent to infringe.

Library of Congress Cataloging-in-Publication Data
A catalog record for this book has been requested

ISBN: 978-0-367-72605-8 (hbk)
ISBN: 978-0-367-72603-4 (pbk)
ISBN: 978-1-003-15551-5 (ebk)

DOI: 10.4324/9781003155515

Typeset in Bembo
by Apex CoVantage, LLC

To Dad, Mam and Mary

CONTENTS

Preface	*viii*
Introduction	1
PART 1	**17**
1 A Conceivable Common Good	21
2 A Critically Uncommon Good	48
PART 2	**67**
3 Planning and the Common Good	69
4 The Planner and the Common Good	87
5 Advancing the Common Good	100
6 Planning for the Common Good	114
References	*120*
Index	*137*

PREFACE

> Doctrines which are supposedly derived from the sober examination of some domain into which the self doesn't and shouldn't normally obtrude actually reflect much more than we realize the ideals that have helped constitute this identity of ours.
>
> (Taylor, 1989: ix)

This book was born in a pandemic. As the world went into lockdown, I hunkered down to reflecting, reading and writing on how we think about planning. Feeling like a stowaway among theorists, I was trying to get somewhere while unclear of the destination. Hold up with my books, papers and a stream of media debate on what we should be doing to stem the virus, I wondered if we are missing something by failing to consider the planner as a moral subject. This brought forth the concept of the 'common good' – just as the idea was being undermined in many polities with tragic consequences. Written against the backdrop of the great pause imposed by COVID-19, this book attempts to first take stock of how the most influential perspectives in planning theory have approached the very 'idea of planning'. From here, the book argues that there is a co-constitutive relationship between planning, the planning subject and the common good. This book doesn't follow the in-vogue path of critical commentary, deconstruction or the new ontologies of much contemporary writing in planning theory. Nor does it offer a roadmap, set of solutions or explosive polemics. Instead, it supplies a novel way of understanding what propels travellers in planning as they endeavour to navigate the challenges of context. What this book presents is a way of seeing from within what we are trying to achieve when we think about, debate and do planning.

No stowaway makes it to the other side without some help. For this, I am grateful to Gerry Clabby for our enjoyable conversations and for agreeing to feature in this book, to Huw Thomas for his unfailing generosity and sagely advice, as well as to Kate Schell, Sean Speers and Jyotsna Gurung for steering me through the process at Routledge.

INTRODUCTION

An Enduring Concern

Appeals to the 'common good' or 'public interest' have long been used to justify planning as an activity[1] (Alexander, 2002b; Alfasi, 2009; Chettiparamb, 2016; Moroni, 2004, 2019; Tait, 2016). Indeed, in its ten-year Corporate Strategy for 2020–2030, the Royal Town Planning Institute stated as its mission, 'To advance the science and art of planning, working for the long-term *common good* and well-being of current and future generations' (RTPI, 2019: 3 – emphasis added). Unsurprisingly, scholarship has expended considerable ink exploring how the common good concept has been advanced to legitimise planning activities conducted by the state (Campbell and Marshall, 2002) and as a norm invoked by practitioners when seeking to ground their activities (Campbell and Marshall, 2000). This research has demonstrated how many of the traditional arguments for planning in the common good are rooted in the perception of it as a 'technical' profession predicated on the possession of design skills and conducted according to a model of rational comprehensive decision-making (Campbell, 2012a; Owens and Cowell, 2011). However, such research also stresses how this linear view of planning has been thoroughly undermined over the past four decades.[2] Accordingly, it is now generally accepted that planning is an inherently political activity, informed by values and often conducted against a backdrop of competing interests and power asymmetries (Campbell, 2012b; Fainstein, 2010; Hendler, 1995; Howe, 1994; Tait and Campbell, 2000; Thomas, 1994). This view has weakened the position of the common good justification for planning. Indeed, today there exists a well-established critical suspicion in planning academia of potentially universalising concepts (Sandercock, 1998). This has resulted in some approaching the idea of the common good with a significant degree of cynicism (Reade, 1997; Sandercock and Dovey, 2002). Others have gone further in suggesting that the idea of a common good or public interest provides little more than 'a flexible construct for the articulation of disparate views' (Grant, 1994: 73), with contending positions exploiting it to support their arguments. Moreover, recent research has empirically demonstrated the difficulty experienced by planning practitioners in identifying what the 'common good' may entail (Tait, 2011), and even the relevance of such a concept to their daily activities (Murphy and Fox-Rogers, 2015).

Nevertheless, the idea of the common good endures if not in name, then certainly in spirit, as prominent theorists make valuable contributions to discussion on how we can ensure that planning operates as an ethically attuned pursuit (Campbell and Marshall, 2000; Flyvbjerg, 2004; Lo Piccolo and Thomas, 2008; Tait, 2016). This is because 'planning is fundamentally an ethical activity, as it raises questions about what should be done, for whom and by whom, and with what benefits and losses' (Watson, 2003: 404). As such, we are left with the unavoidable question as to how an ethically sensitive common good should be understood. This book tackles this issue by proposing that the common good should not be conceived as something pre-existing and 'out there' to be identified and applied, or something simply produced through the correct configuration of democracy. Instead, it is contended that the common good must be perceived as something 'in here', which is known by engagement with the complexities of a context through employing the interpretive tools supplied to one by the moral dimensions of the life in which one is inevitably embedded.

Planning's Public Interest/Common Good

Theoretical Typologies

'Normatively, planning is premised on the inherently hopeful conviction that a better future is possible than would have occurred in the absence of "planned" intervention' (Campbell et al., 2014: 45). In this sense, planning is an intrinsically moral activity propelled by a desire to make the world a better place. Yet, what constitutes 'better' is frequently contested such that the planner must make decisions that privilege certain conceptions of 'better' or at least facilitate compromise on what 'better' may mean. This necessitates the deployment of some kind of rationale for the decision taken, a choice that has an inevitable moral-political dimension 'in a climate where appeals to a planner's scientific expertise are no longer seen as valid' (Tait, 2011: 158). Consequently, planning theory has wrestled with attempts to effectively profile the ways in which the public interest can be reasoned. For the most part, these efforts have drawn on political theory or reformulated ethical theory in political terms when seeking to furnish typologies that categorise the different logics proposed by others for determining the public interest. This is often then used as a platform on which to build the author's specific contribution via a new abstraction on how the public interest should be conceived. An early example of this in planning theory was provided by Edward Banfield (1973 (1959)), who develop a fivefold 'conceptual scheme' derived from a review of ethical philosophies. He nuanced the differences between the utilitarianism of Bentham (2004, 2019) and Mill (2015) to identify a developing trajectory towards increasingly more subtle forms of individualism. This was then compared with 'unitary concepts' of the public interest drawn from communitarian political philosophy that focused on issues seen to be of common concern to a community of evaluators. More an early adumbration of how the public interest may be conceived in planning than a proposal for thinking anew, this schema nevertheless provided the basis upon which Elizabeth Howe built her seminal typology (1992). She combined the work of Banfield with that of Held (1970) to identify the importance of delineating between the respective subjective and objective ethical assumptions of utilitarianism and unitary conceptions. Through a detailed argument, she concluded that the planner's 'ethical obligation to be responsive and responsible to public values' (Howe, 1992: 230) does not mean that it is 'somehow improper to raise and argue for values and goals' (p. 244). What was required was clarity in how those values and goals are being argued for in the public interest.

Referencing Howe's work, but with a more profoundly philosophical consideration of how the concept of 'interests' has evolved in the modern period, Campbell and Marshall (2002) trace the ways in which the 'public interest' has been regarded in planning literature throughout the latter half of the twentieth century. This survey leads them to conclude that the dominant view within the planning academy is 'dismissive of the public interest because it is either too vague to be useful or because it is an elitist and potentially undemocratic idea' (p. 173). However, through a valuable review of the normative interpretations of the public interest in planning, Campbell and Marshall demonstrate that 'although a contentious concept, the idea of the public interest has never entirely been abandoned,' inferring that 'different conceptions have held sway at different times' (p. 173). Their argument builds to a typology that distinguishes deontological (procedurally focused) and consequentialist (outcome-focused) conceptualisations of the public interest in planning. An important dimension of this work is the expansion of Howe's subjective-objective dichotomy to include an 'intersubjective' understanding of the public interest in deontological conceptualisations when referencing dialogic planning theories, such as communicative planning (Forester, 1999a; Healey, 2006; Innes, 1996). This in-depth survey leads Campbell and Marshall to conclude that although the idea of the public interest is often 'scorned' by planning theorists, 'it nevertheless remains the pivot around which debates about the nature of planning and its purposes turn' (2002: 181). Indeed, running counter to the tide of planning theory at the turn of the twenty-first century, they conclude their paper by arguing against the prevalence of procedurally focused dialogic theories as a means for determining planning's public interest by asserting that 'no account of planning, politics and the public can be of value if it is empty of all substantive content, of what is at stake' (p. 182).

While referencing the work of Campbell and Marshall (2002), Earnest Alexander (2002b) provides an alternative typology. He opens his argument by contending that the public interest concept serves three roles in planning – namely, to legitimise the discipline, to supply a norm for practice and to provide a criterion for evaluating planning policies, projects and plans (Alexander, 2002a). He then traces the origins and evolution of the concept from classical to contemporary times to infer that a substantive public interest criterion is required for plan evaluation. Alexander's schema draws on ethical, legal and constitutional scholarship to sketch how the public interest is broadly conceived with either a substantive or procedural focus.[3] He further nuances this simple division into a fourfold classification of aggregative, unitary, deontic and dialogical approaches. Similar to Howe, these are then related to 'objective' and 'subjective' perspectives and associated with different theoretical perspectives ranging from utilitarianism and individualism to communitarianism and communicative practice. As with Campbell and Marshall (2002), Alexander also includes an 'intersubjective' perspective on how the public interest is determined. Alexander's divisions are both subtle and complex. While it is not necessary here to stray into a tangential dissection of his complicated typology, his interpretation of an 'intersubjective' perspective requires some unpacking. In particular, Alexander's sensitivity to the intersubjectivity of how norms are constituted, communicated and institutionalised leads him to conclude that 'unitary concepts of the public interest are the most widespread (though rarely recognised)' (Alexander, 2002b: 240). For him, such unitary concepts differ from utilitarianism. Instead of attempting to aggregate individual values or adjudicate on such aggregations as different forms of utilitarianism do, 'unitary concepts base the public interest on some collective moral imperative that transcends particular or private interest' (p. 230). Hence, unitary concepts of the public interest are substantive. However, he holds that the complexity of planning's political embedding 'makes a simple unitary public interest principle inconceivable, and its substantive application infeasible' (p. 241). Alexander's innovation is therefore to note, 'In any application of unitary concepts of the public interest, what emerges is some recursive combination of dialogic and deontic approaches' (p. 232), where 'deontic' is seen to mean rule or norm-based action. In this way, he proposes that the substantive and the procedural are mutually constitutive in determining understandings of the public interest in planning. Specifically, he argues that while the dialogical concept of the public interest (e.g. communicative planning) may be useful as a 'kind of default legitimation for planning,' it nevertheless 'fails to provide the substantive content that planners need if the public interest is to have any value as an evaluative criterion for plans and planners' decisions' (p. 234). Hence, he reasons, 'Discourse (ideological, political and technical) develops and adopts normative systems (norms, laws, rules, policies, plans) that provide the deontic framework for the disposition of planning issues, and the discretionary arena for further (political and technical) dialog' (p. 240). In this way, he echoes Campbell and Marshall by rejecting growing appeals to Habermasian inspired conceptions of the public interest as an unrealisable planning response to the cold calculations of utilitarianism or the limited moral horizons of communitarianism.

The ongoing 'pivotal' nature of the public interest to planning that was identified by Campbell and Marshall (2002) at the turn of the twentieth-first century is confirmed by the prominence given to Stefano Moroni's chapter on this topic in Routledge's *Handbook of Planning Theory* (2018) published almost two decades later. Here Moroni adopts a novel approach to discussing 'the fall from grace of the idea of the public interest' by examining four different arguments for its 'inexistence' – namely, that it does not exist *as a fact*, that it does not exist *as an a priori criterion*, that it does not exist *as an extra-individual value* and that it does not exist *as an always overriding substantive value*. Moroni undertakes an erudite summary of political theory in teasing apart the differences between these arguments to identify their empirical, ethical and meta-ethical grounding. These are related to perspectives he terms 'divergentism' (an overwhelming divergence of interests that inhibits a shared sense of the public interest), 'dialogical proceduralism,' 'classical liberalism' and 'value-pluralism'. He then uses this review as a foundation upon which to advance his own public interest concept of 'nomocracy' rooted in the libertarian thinking of Hayek (1978, 1982), which he has developed and elaborated across a series of publications (Alexander et al., 2012; Moroni, 2010, 2014).[4] While Moroni concludes with a concept of the public interest which differs from that of Howe, Alexander or Campbell and Marshall, his review nevertheless echoes their belief that 'one of the main tasks of planning and political theory is to rethink the public interest' (Moroni, 2018: 78). Angelique Chettiparamb (2016) furnishes one such means for rethinking the public interest. In contrast to Moroni, her review of the public interest concept is largely confined to the concept's manifestation in planning literature. However, in keeping with the broad approach of Moroni, she examines the reasons why the concept has been repeatedly dismissed by planning academia as a base from which to explore its persistent, if sometimes only implicit, relevance. Chettiparamb's mobilisation of the public interest concept strongly associates it with the idea of 'social justice'. As she affirms,

> The question of public interest for planning is not far removed from the question of social justice given that allocation, be it of land, resources, funds, energy, social goods, etc. is central to the act of planning . . . The logic of public interest thus coincides with the logic of public values including those of social justice.
>
> (Chettiparamb, 2016: 1287)

Following this association, Chettiparamb undertakes a discerning review of social justice theorists who have had an impact on planning theory discourses regarding the public interest. In this, she narrates a dialectic between Rawls (1971), Sandel (1984), Walzer (1983), Nussbaum (2008), Young (2003) and Fraser and Honneth (2003) to emphasise the challenge in locating 'how or what principles can be used to articulate a system that has the capacity to be responsive to relative meanings while at the same time retaining the capacity for adjudication' (Chettiparamb, 2016: 1290). Her response is to build upon the foregoing review and mobilise the 'complexity theory' she has developed and explained across numerous papers (Chettiparamb, 2006a, 2006b, 2013, 2014, 2019). Specifically, she extends our understanding of the public interest by using a case study of an Indian poverty alleviation programme to demonstrate how in complex systems 'notions of the public interest get differentiated in scale' (Chettiparamb, 2016: 1300), such that the principles governing how the public interest is identified may differ at different scales of governance, yet do not necessarily conflict with how it is identified at other scales. Chettiparamb argues that this is because complex systems of governance 'are necessarily organised through scales (not hierarchies)' (p. 1300). Her work thereby helps bridge the theoretical and the empirical, adding nuance to both.

Empirical Research

Several researchers have pursued empirically focused accounts of the common good or public interest in planning (Howe, 1994; Low, 1991b; Meyerson and Banfield, 1955). Although of high quality, much of this work is somewhat dated, as it emanates from a period prior to the significant expansion of theoretical diversity in planning academia that blossomed in the closing decade of the twentieth century.[5] However, notable contributions have been made by contemporary researchers. Among such recent literature is Tait's (2011) work examining the links between 'trust' and the public interest in the micropolitics of planning practice. Employing an ethnographic methodology to investigate the embedded nature of trust relationships in a planning office, Tait identified two discrete articulations of public value, 'each associated with a different conceptualization of the "public interest", and each promoted by a different group with different values, different priorities, and different ideological commitments.' As he notes, 'These political differences percolated through the interpersonal relationships observed' (2011: 165). Centred on a distinctly spatialised 'public realm' conception of the public interest and a separate economised 'growth' agenda, these different interpretations forged alternative networks employing distinct discourses that framed the world in different ways such that the legitimacy of evaluations and interventions were dissimilarly conceived. Tait concludes that

> it is only by critically examining the underpinning rationale behind these formulations and offering intellectually and practically coherent answers to questions of the relationship between the individual and the collective that planning can become a system that is comprehensively trusted.
>
> *(2011: 168)*

Murphy and Fox-Rogers (2015) have subsequently expanded our appreciation of practitioner perceptions of the common good through a series of semi-structured interviews with twenty urban planners working across four Irish planning authorities within the Greater Dublin Area. Through a detailed dissection of viewpoints, these authors conclude that the most striking aspect of their results 'is the extent to which planners' ideas of the common good are bound up with pluralistic notions of collaborative planning where the common good is about refereeing between stakeholders rather than attempting to act more radically to transform power relations' (p. 240). Interwoven with a critique of neoliberalism and in reference to Davidoff's (1965) planning polemic, Murphy and Fox-Rogers infer that 'a very good argument exists for advocacy planning' as 'planners have a moral obligation to work in the interests of marginal and disempowered stakeholders' (2015: 240). Schoneboom et al. (2020: 464) draw similar conclusions in their research, contending that 'perhaps the truly radical act would be for the profession to recover the political sensibility that underlies its *raison d'etre* (sic) and mobilise to reverse these damaging trends.'

Taking a different approach, Christopher Maidment's (2016) investigation into how the public interest is variously conceived in planning England's Peak District National Park initially adopts the typology advanced by Campbell and Marshall (1999, 2002) discussed previously. This is interlaced with Dewey's (1954) work on democratic representation to conceive interests relative to publics that only come into existence 'when the consequences are indirect and therefore cannot be dealt with through the participation in making of the decision by all of those affected' (Maidment, 2016: 369). Resonant with the conclusions of Chettiparamb (2016), albeit without the theoretical machinery of complexity theory, Maidment surmises the importance of attending to scale as a way of recognising when different conceptualisations of the public interest might be practically constituted by multiple publics. Hence,

> The concluding argument is not for either a singular or a dialogical approach to the public interest but is that planning must pay greater attention to who, and what, is included in the multiple publics whose interests it must address.
> *(2016: 384)*

These readings are echoed in the conclusions drawn by Puustinen et al. (2017) in their analysis of Finnish planners. Deploying a framework formulated from a review of political theory that cross-references individual and collective based concepts with regulation and non-regulation focused perspectives, these researchers conjecture,

> The co-existence of various public interest conceptions can make the concept a dubious rhetorical tool in planning practice. Without the explication of the discursive context, the concept is devoid of meaning. Thus, when truly seeking justification to (sic) planning decisions by appealing to public interest (sic), its discursive context has to be explicated.

The authors infer that without such discursive context, 'the concept could be used to promote contradictory normative purposes' (p. 93). Research by Searle and Legacy (2021) into the planning of Australian transport infrastructure partially confirms this conclusion by reaffirming the danger of the public interest label being used as a rhetorical justification that is normatively confused. In particular, they echo concerns expressed by Murphy and Fox-Rogers (2015) that such confusion can result in the public interest concept being 'co-opted to legitimise market rationality and economic values over values of justice, equitability and environmental sustainability' (Searle and Legacy, 2021: 841). However, they hold hope that 'planning's inherent logic of public purpose can be used by varied groups aiming to dislodge planning from narrow and path-dependent economic logics' (p. 841).

Thus, a number of theoretical and empirical works specifically examining the concept of the public interest/common good are available in the planning archive, each characterised by high-quality research that mines political and ethical philosophies and/or reflectively evaluates the results of empirical investigations. However, the theoretical work is largely confined to producing typologies that walk a well-worn path of debate borrowed from political science and ethical philosophy to furnish a foundation upon which to construct an abstracted account of the public interest tenuously referenced to practice. Conversely, much of the empirical work seeks to probe understandings and uses of the public interest or common good concepts without adequate theoretical attention to how such uses are contextually interpreted in surrogate forms that give meaning to action which exceeds simply instantiating neoliberal logics. Maidment (2016) has somewhat helped address this deficit through his appreciation of multiple 'publics' styled after Dewey, while Chettiparamb's (2016) sensitivity to agency in her mobilisation of complexity theory (Chettiparamb, 2019) adds nuance to our understanding of how different communities can reference dissimilar ethical registers in justifying their assessment of the public interest. Whereas such work adheres to the belief that planning is legitimated by appeal to the public interest/common good, its elucidatory potential is nevertheless hindered by theoretical lacuna on how planning and the common good may mutually profile each other. Thus, this book argues for reconceiving this relationship as co-constitutive. That said, it is contended that a fully rounded appreciation of this co-constitutive relationship is lacking without an understanding of the 'planning subject' as the morally informed decision-maker who carries significant responsibility for deciding how the common good should be determined.

The 'Planning Subject'

There is now a bank of empirical work exploring the ethical challenges planners may confront in their practice (Fox-Rogers and Murphy, 2016; Howe and Kaufman, 1979; Lauria and Long, 2017; Loh and Arroyo, 2017; Loh and Norton, 2013; Othengrafen and Levin-Keitel, 2019). Much of this research focuses on using surveys and interviews to identify the roles planners assume and the behaviours they espouse. This work usually seeks to 'measure' the use of different perceptions regarding correct ethical action and highlight how practitioners 'struggle with emotional and ethical dissonance in their attempts to balance their private ethics, their workplace norms and culture, and their professional code of ethics' (Lauria and Long, 2019: 393). Another strand of work seeks to examine how planners negotiate the institutional constraints of their position. Exemplary research in this vein includes the now classic work by Thomas and Healey (1991) on the ethical dilemmas experienced by practitioners that seeks to give voice to planners wrestling with the implications of their everyday activities, as well as more recent research by Andy Inch (2010), which illustrates how English planners have sought to negotiate attempts by managerialist agendas to regulate their identities through adopting practices that throw into question the normative promises of touted planning reform. Research by Ben Clifford and Mark Tewdwr-Jones (2014) into the government reforms introduced by New Labour in Britain between 2004 and 2008 echo Inch's findings through an institutionally sensitive reading of how planners' activities on the frontline of practice may undermine the objectives that the reforms sought to achieve. Here planners are shown to be active participants in the interpretation of change programmes in ways that neutralise, transform or selectively instantiate reforms through the professional and organisational cultures in which they are embedded.

Yet, perhaps consequent on the focus of this work, a theoretical conception of the 'planning subject' is somewhat absent from much of this empirical research. In this context, Beauregard (1998) usefully highlights how theorists are not simply required to detail the activities of practice but must also 'write the planner' who will perform these actions. Through an exploration of the respective 'planner' conceived in John Forester's (1982) 'communicative action' and Leonie Sandercock's (1995) 'postmodernism,' Beauregard stresses that theorists are compelled to address the identity of the planning subject they envisage if their theoretical proposition is to provide clarity on how it confronts planning theory dilemmas. Nonetheless, rather than proposing a clarifying conceptual model, Beauregard highlights an issue requiring redress. Zanotto (2019) helps tackle this deficit through her ethnographic research into how Brazilian private-sector planners manage the potential regressive outcomes of their professional activities. Theorising that 'planners (both in public and private sectors) may often feel torn between what planning scholarship considers progressive and what developers deem feasible and desirable' (p. 49), she hypothesises processes of political, professional and valuative 'detachment' as 'coping mechanisms' used by developer-enabling planners to restrict ethical self-reflection on their practice. These are employed to help them eliminate the potential guilt associated with acting against their personal beliefs (p. 48). Although a useful contribution to profiling the moral planning subject, the focus of this research on 'planning consultants' leaves us with only a partial understanding of 'the planner'. Against this backdrop, theoretical work by Michael Gunder and Jean Hillier (2009) supplies perhaps the most sophisticated conceptualisation of the 'planning subject' in the academic literature. Drawing on Jacques Lacan's psychoanalytical critical social theory and heavily influenced by the writings of Slavoj Žižek (1989, 1997), they construct a post-structuralist reading of planning and the planner that seeks to destabilise long-held truisms of practice. Centred around a series of linked concepts grounded in a 'desire' to satisfy an inherent 'lack' in our relations with the world and those in it, their complicated theory construes the planning subject in a way that facilitates comprehension of how the planner's consciousness is shaped by planning education (Gunder, 2004), and subsequently how the expectations of professional practice shapes norms and values (Gunder and Hillier, 2004). However, their critical theory allows little space for the 'common good,' except as a 'fantasy' born of a yearning 'to cover the antagonisms and inconsistencies that pervade the social field' (Dean, 2001: 627; quoted in Gunder and Hillier, 2009: 191). In their view, the common good would simply serve as a 'master signifier' that ties 'together multifaceted and often muddled and conflicting arrays of narratives under one universal and iconic signifier' so as to allow 'shared and harmonious social identifications' (Gunder and Hillier, 2009: 16). Indeed, their work is critical of the very constitution of planning as an enterprise of worldly intervention founded on imaginings of what ought to be. Moreover, Gunder and Hillier do not (nor claim to) offer a positive path for planning out of their critical interpretation. In this sense, it is notable that much of Hillier's work has since focused on attempts to outline a non-essentialist and relational theory of planning grounded in the works of Gilles Deleuze and Félix Guattari (Hillier, 2011, 2015, 2018). While appreciating the subtly of their approach, its relevance to the objectives of this book is foreclosed by the post-structuralist nature of the critical-theoretical perspective they adopt. Accordingly, a new theoretical approach is required.

Theoretical Approach

The broad contours of this book follow a line inspired by the work of the philosopher Alasdair MacIntyre. Although enjoying a reasonably high profile in political and moral philosophy, the potential of MacIntyre's work is largely unexplored in the context of planning (Lennon, 2015b, 2017; McClymont, 2019; Throgmorton, 1996). In summary, he rejects the concept of an independent 'subject' fashioned by the modern liberal individualism of Kant and Mill that has proved so influential in contemporary moral and political philosophy, and which has helped shape recent thinking in planning scholarship via appeals to the work of luminaries such as John Rawls and Jürgen Habermas. However, MacIntyre also submits a withering critique of Marx's argument against liberalism, finding that it evacuates agency from the reasoning 'subject' by creating an untenable structuralism. He thereby finds himself agreeing with Nietzsche that moralities born of such philosophising essentially mask a 'will to power' wherein ethical theories of apparent objectivity at best disguise implicit biases and at worst can be abused for conscious manipulation. Nevertheless, he does not support the Nietzschean rejection of all forms of moral philosophy, as he believes this leads to an arbitrariness that is unreflective of how people actually live their lives. In this vein, he repudiates postmodern contentions on the priority of subjectivity that helped reorientate much planning theory in the closing decades of the twentieth century. Hence, MacIntyre attempts to stride a path between flawed appeals to ethical objectivity and nihilistic subjectivity. He achieves this by focusing his attention on 'the conditions that enable human agents to recognize and choose what is good and best, for themselves and others, as members of the communities to which they belong' (Lutz, 2012: 2). This path takes

MacIntyre on a journey through moral and political philosophy where he identifies the inadequacies of both in conceiving the thinking subject. Ultimately, he arrives at the conclusion that such philosophising requires renewal via a return to studying how moral agency is intersubjectively profiled by a community in its way of thinking about and acting in the world.

Complementing but not duplicating the insights of MacIntyre is the work of the moral and political philosopher Charles Taylor. Even more so than MacIntyre, Taylor's work has been largely overlooked by planning academia, perhaps resultant from his misgivings about the procedural ethics that have been so influential in planning theory since the 1990s. Indeed, like MacIntyre, Taylor is suspicious of how modern moral philosophy since the Enlightenment has sought to emulate the natural sciences by attempting to establish objective standards against which matters of good/bad, right/wrong and better/worse can be neutrally determined. Drawing upon a pantheon of philosophers such as Aristotle, Hegel and Heidegger, as well as upon MacIntyre, Taylor launches a project in 'philosophical anthropology' (Taylor, 1985a, 1985b) that seeks to distinguish the transcendental dimensions of human experience. Most pertinent to the objectives of this book, he develops a phenomenology of moral experience. His work thereby helps bridge the abstractions of MacIntyre with the particularities of how people negotiate what Taylor refers to as the 'moral spaces' we each encounter in our daily lives. He achieves this by adopting a two-pronged strategy. First, he reasons that people are required to account for reality through the experience of it, and in doing so formulate ways of giving meaning to it. This meaning-making process is bidirectional, as a dialogue is established through 'webs of interlocution' (Taylor, 1989: 36) wherein people intersubjectively construct a moral landscape that both transforms and is transformed by interaction with the world. Intersubjectivity is here understood to indicate the condition whereby the meanings of places, objects, states, beliefs, etc., and their respective relationships are constituted and shared by people in their interactions. These meanings are employed as a resource to interpret and navigate life. Intersubjectivity thereby refers to how people understand the world around them as beings whose perspectives are always mediated by their location in a world of others. Taylor's second move is to demonstrate how the moral horizons we intersubjectively create are dynamic rather than static. He accomplishes this through a philosophical history of morality. However, 'when doing philosophical history, Taylor, like Foucault before him, is interested not so much in general truths about the human condition as in contingent constellations of human self-understanding, specifically those that prevail in the modern world' (Smith, 2002: 8). In essence, Taylor adopts both an ontological and an historicist perspective to derive a theory of the moral self that responds to the mistakes he sees made by approaches which seek to formulate an objective ethics for determining what ought to be done. He undertakes this by reinvesting the self with a sense of moral agency.

Thus, by tracing lines of thought through different philosophical foci, this book engages a novel integration of aligned ontological approaches to supply an original theoretical frame for understanding how planning's common good is conceived.

Nevertheless, the book does not endeavour to interpret MacIntyre in the terms of Taylor. Instead, each theorist is consulted to furnish the conceptual tools necessary to unlock the specific issue at hand. As such, the theorists enter the conversation as complements to each other rather than in competition. The use of each is carefully chosen to build upon the conceptual programme provided by the other such that the order of their appearance in this book reflects a different level of focus as we progressively reformulate our understanding of the relationship between planning, the planner and the common good. Aided by these theorists, this book seeks to go beyond simply outlining a theory for identifying the common good 'sought after' by planning. Instead, it endeavours to chart new terrain for exploring how the very nature of planning is co-constitutive with the concept of the common good it espouses, such that one profiles the other in an ongoing dance through time and context. The book advances the view that it is necessary to conceive planning as a moral enterprise to understand how the common good is appreciated in practice. Hence, the book avoids and at times argues against the deployment of theoretical perspectives that predetermine what the common good means and/or how it 'should' be identified. As an alternative, it is contended that a more subtle understanding of how the 'moral subject' relates to the intersubjectively received and (re)interpreted norms available to them in decision-making is central to understanding how the common good is understood and how planning is contoured as an activity responding to this understanding.

Outline of the Argument

This book argues that it is vital to consider people as existing in what Arendt referred to as 'webs of human relationships' among a plurality of unique selves (1958: 233), a point concisely described by Young as 'togetherness in difference' (1993: 130). It is also agreed with Campbell that 'There is . . . no formula available through which the right and the good may be calculated,' and as such, 'planners are fundamentally concerned with making *ethical judgements*' (2006: 92 – emphasis in original). Specifically, this book examines the relationship between planning, the planner and the common good in a way that shows how each is co-constitutive through a dialogue that is cast by how we understand what is choice worthy and what is a worthy choice. The structure of the book is mapped out to take the reader on a journey from the abstractions of broad philosophical arguments to the frontline of planning. Hence, the book is primarily conceived as an extended theoretical argument rather than an empirical discussion. Nevertheless, both real and hypothetical examples are referenced throughout the book to help ground the argument in relatable experience. Following this introduction, the reader begins their voyage with a review of planning theory.

Planning theory has multiple purposes. Some theories describe how planning works. Others prescribe what planning should do. More provide powerful critiques that imply possible alternatives. Then there are those that are a mix of these and further agendas. Yet as contended in Part 1, there is broad-based interest in using the push and pull of debate to advance the metier. Much of this debate centres on the accuracy of representation that in the rough and tumble of argument frequently focuses on issues of power and the associated topics of privilege, marginalisation, inclusion and exclusion. Some recent work seeks to gain a foothold among prominent contributions by taking theory in a different direction through rethinking the ontology of planning with a greater focus on clarifying what planning 'is' rather than what planning 'should do'. Nevertheless, the propellant to better understand what is done by whom, how, when, where, why and with what consequences remains a line threading the field together. It is therefore curious to note that, notwithstanding their differences, most established positions in planning theory share certain blind spots that curtail our ability to appreciate the very impulse for action and the bridge that connects theory with the world. Accordingly, the two chapters of Part 1 seek to identify and discuss these blind spots with a view to addressing them in subsequent chapters.

The first of the two chapters in Part 1 (Chapter 1) surveys those influential theoretical perspectives that share an implicitly or explicitly focused approach to the identification of the common good. The chapter begins by reviewing endeavours to render planning a scientific activity and how this provoked the emergence of an influential line of theory concerned with fostering more inclusive deliberation. Although the approaches reviewed in this chapter may initially appear contradistinctive, it is argued that they share a presupposition that the common good should be an issue of concern that is possible to identify. The theoretical perspectives reviewed in Chapter 2 take a contrasting position. These fundamentally spurn the idea of a 'common' good, viewing it as at best an outmoded concept and more frequently as one that impedes the expression of difference. It is argued that the explanatory potential of each perspective reviewed in this chapter is stymied by either its explicit suspicion on claims to the 'commonness' of any good or the difficulties in rendering its concept of the subject operable in a planning context. What Part 1 demonstrates is that almost all planning theories either explicitly affirm or are implicitly forged by a normative impetus to help make the world 'better,' even if this is simply to 'understand better' what planning is and how it operates. As such theories are inherently abstractions, what they direct attention towards is a way of conceiving the world that facilitates the specification of how challenges encountered can in some way be addressed for the benefit of society in general. Thus, most planning theories possess an intrinsic sense of the 'common good,' even if this is only tacitly indicated in what it is they oppose, including the very idea of the common good itself (Sandercock, 1998). This is because planning as a practice of decision-making is legitimated on the basis that it serves the common good (Alexander, 2002b). To deny the common good as a grounding for planning is thereby to deny its raison d'être. The work of Paul Ricoeur (1995) is employed in the conclusion of Chapter 2 to show how understanding what threads planning together as a discipline involves appreciating the moral commitment it holds to advancing the common good, albeit that such a common good may be variously interpreted across space and time. This thereby shows how planning, and the common good that it seeks to advance, exist in a mutually constitutive relationship that gives planning its raison d'être and the common good its meaning. However, understanding how the common good and planning shape each other involves a grasp of how planning thought and action are contextually embedded.

Yet, academics and those in practice are well aware that it is not easy to supply a simple explanation of what planning is or what planners do. Indeed, questioning planning's identity may provoke puzzlement when a definition is sought but a succinct response is unforthcoming. As noted by Vigar (2012: 374) when referencing MacDonald (1995: 201), 'Planning shares with professions such as accountancy a basis in an 'esoteric collection of areas of knowledge', rather than a basis in esoteric knowledge.' While it is true that certain technical competencies are required to do planning, what such competencies are may vary widely depending on the context and concern of the planning activity engaged in. Nevertheless, as planning academics and practitioners, we intuitively feel that there is an intrinsic unity to our discipline, even if from the outside it can at times appear as little more than a collection of different activities operating at different scales, with different foci and methods, where those in different arms of the profession appear to speak different languages. How after all can the transport planner wielding the tools and jargon of geographical information systems be positioned within the same discipline as the planner interpreting the aesthetic impact of a development proposal in a scenic rural landscape? A fruitful response might be to think less on 'what' planning does and refocus attention onto 'why' it does what it does. This involves identifying a commonality within the expansive family of theory and practice accommodated beneath the 'planning' rubric. Hence, Part 2 argues that understanding what planning is must remain sensitive to the moral impetus lacing together the variety of planning thought and activities across space and time. It is contended that no matter what the theoretical school or practice family, planning's identity is forged by a perpetual effort to do what it does for the common good. Although the common good may be conceptualised in varying ways, thereby influencing how planning manifests differently in thought and action in different contexts, this does not reduce the central character-defining identity of planning as an activity directed at advancing the common good. This is because issues of better/worse and good/bad concern the very 'soul of planning' as an activity 'premised on the expectation that through intervention and action, better space and placed-based outcomes can be achieved than would otherwise have been the case' (Campbell, 2012a: 393). Therefore, 'planning is in many fundamental ways a series of statements about what we take to be right or wrong and what we take to represent the highest priorities of the society in which the planning is undertaken' (Wachs, 1995: xiv). However, there is no simple blueprint for discerning the right and the good in a world of 'wicked problems' (Rittel and Webber, 1973) filled with 'inherent uncertainty, complexity and inevitable normativity' (Hartmann, 2012: 242). Rather, there are only different, and sometimes contending approaches to identifying what *ought* to be done in the context of what *can* be done. The endeavour to locate a means to inform this normative aspect of planning has a long history and a short past, wherein there is an extensive convention of justifying governmental action in the name of the 'common good,' while concerted effort to systematically examine the ethical dimensions of planning dates back just a few decades (Campbell and Marshall, 2002; Hendler, 1995; Wachs, 1985). Yet in a world where 'knotty questions abound' (Kaufman, 1981: 196) on 'what' *should* be done and 'why,' tackling the thorny issue of planning's relationship with the common good has become an oddly uncommon topic in contemporary theoretical debates than one might expect for a discipline whose raison d'être is predicated on its delivery. Hence, building on the review of Part 2, Chapters 3 and 4 set out to profile the relationship between planning, the planner and the common good. Specifically, Chapter 3 borrows a series of philosophical insights from the philosophy of Alasdair MacIntyre to facilitate an appreciation of the 'situatedness' of planning. This results from the inherent nature of planning as that which seeks to advance the common good. It is this endeavour that gives planning shape and the common good significance, even where this is not explicitly acknowledged. The argument advanced is that this phenomenon is consequent on how the values carried by the 'moral subject' of planning decision-making (i.e. the planner) are shaped through both their relationships with others and their relationship with the 'tradition of planning' that extends as a form of reflective deliberation across time and space.

Building upon this line of argument, Chapter 4 mines the work of Charles Taylor to extend the MacIntyrean insights of Chapter 3 in exploring how one's 'situatedness' helps shape perceptions and actions in ways that are imbued with moral perspectives, which are given representation in appeals to ethical standards. Whereas Taylor frequently uses the words 'moral' and 'ethical' interchangeably, these words are treated here in their more conventional academic form such that 'moral' signifies personal views along a spectrum of what is right and wrong, while 'ethics' is used to indicate those views held by a community of interpreters. In this sense, 'ethical' denotes professional conduct, while 'moral' indicates one's beliefs on qualitative distinctions between good/bad, right/wrong, better/worse, etc. Nonetheless, as argued in this chapter, morals and ethics interact such that one influences the other. Taylor's contention for the validity of a 'moral realism' as a perspective for explaining what motivates people to think and act is both explained and adopted here as a means to examine the limited moral pluralism fashioning the ethical dimensions of one's 'situatedness'. At the heart of this chapter is an investigation of how 'moral frameworks' are formed and shared in ways that influence thought and action in planning.

Chapter 5 supplies an illustration of these insights. This is achieved through an examination of the emergence, evolution and institutionalisation of the 'green infrastructure' approach in Ireland. It provides detail on how conventional planning was transformed by reflection on what constitutes the common good. The case study illustrates how change can be motivated by a desire to move planning towards a position of greater practice excellence in representing the common good. As such, it shows how reflection on what constitutes better planning – and what instantiates it in practice – is guided by commitment to a sense of the common good that is determined by the qualitative distinctions operative within moral frameworks, rather than external to them.

Chapter 6 brings the book to a conclusion by arguing that reasoning in planning on what to do is inherently substantive in nature. It is contended that the moral realism of a limited moral pluralism grounds planning as an activity. To advance the common good from within such a limited moral pluralism constitutes the 'practice of planning' that carries it forward as a tradition directed towards advancing the common good. The chapter closes by bringing the argument full circle back to how the mutually constitutive relationship that gives planning its raison d'être and the common good its meaning is conceived through a narrative understanding extending through time that contours the moral subject of planning as it simultaneously profiles the ethical orientation of the discipline.

Notes

1 The expressions 'common good' and 'public interest' are used interchangeably in the planning literature. However, as the 'common good' is the phrase most frequently used in the moral philosophies informing my thesis, I will generally confine my use to this phrase when outlining my own take on the topic. Nevertheless, I mean my use of the phrase 'common good' to also encompass the 'public interest' as the phrase most often used in planning theory and practice. Accordingly, I will refer to the 'public interest' when reviewing the theories of others who employ this phrase so as to preserve their 'voice'. In this context, it is noted that the frequency of use of the phrase 'public interest' by planning theorists and practitioners indicates how contemporary thinking in this field has primarily drawn upon 'political' rather than 'moral' philosophy in discourse on the subject (e.g. Madison, Rawls, Habermas) or couched moral philosophies in political terms (e.g. Mill's utilitarianism or Kantian deontology). Indeed, my decision to employ the phrase 'common good' instead of 'public interest' reflects the alternative path I take in conceiving what the concept means in the context of planning. In this sense, it is interesting to note that Flathman identifies the ascendancy of the term 'public interest' as carrying individualist and subjective connotations, which until twentieth-century ethical theory were not associated with the concept of 'good'. He concludes, 'It was not until the satisfaction of subjective, even idiosyncratic, individual interests came to be considered a prime object of politics that "interest" could replace "good" as the primary concept of political life' (1966: 14).
2 Some criticism of rational comprehensive planning pre-dates the mid-1970s. However, it was not until this period that acknowledging the political nature of planning gathered momentum in academia.
3 Substantive-focused theories advocate normative ethical principles and judgements to be applied in the evaluation of right/wrong and good/bad with respect to specific actions, plans, policies, institutional arrangements, etc. Procedurally focused theories are primarily concerned with the correctness of the process followed in determining and justifying substantive principles and arriving at decisions. See Harper and Stein (1992). However, as noted by Innes and Booher (2015) and echoed by Moroni (2019), planning theory has often employed an excessively dichotomous distinction between substantive outcomes and decision processes.
4 Moroni's conclusion of what constitutes the public interest in planning echoes that advanced by Nigel Taylor (1994), albeit Moroni's philosophical justifications are different to those provided by Taylor.
5 As outlined in Chapters 1 and 2, it is obvious that several contending theoretical positions have a long pedigree extending back several decades. However, many of these only gained widespread traction in planning theory debates during the 1990s in the wake of postmodern critiques that had been sweeping through the social sciences since the early 1980s. See Allmendinger (2002) and Mandelbaum et al. (1996).

PART 1

> All theory is to greater or lesser degrees normative.
>
> *(Allmendinger, 2017: 17)*

Anyone who has ever dipped their toe into planning theory soon realises that it is a flow of crosscurrents that at different times blend, contend or simply ignore each other. Consequently, any hope of an all-inclusive review of this torrent is doomed to failure. Part 1 thereby holds no pretence to an all-encompassing consideration of the topic. Hence, the following discussion is limited to debates related to the focus of this book – namely, planning in modern western democracies. It is for this reason that excellent work in the fields of postcolonial theory (Bishop et al., 2013; Jackson et al., 2017; Porter, 2016), insurgency in planning (Meth, 2010; Miraftab, 2009) and emerging approaches to informal settlements are not addressed (Amin and Cirolia, 2018; Roy, 2009). Moreover, aware that the ebb and flow of debate throws a variety of different perspectives onto the shores of academic attention, the two chapters that follow focus primarily on those theoretical approaches that have demonstrated a certain fixity against the tide of time. This is undertaken to balance the requirements of conciseness, comprehensiveness and clarity. Accordingly, more recent approaches that have yet to achieve broad purchase beyond their principal authors do not receive detailed attention. It is stressed that this is not an appraisal on the merits of such perspectives, as their currency across time will be the ultimate arbiter of their contribution. Rather, it is simply a matter of practicality. In this sense, provocative planning theory mining the oeuvre of Lacan (Gunder, 2016; Gunder and Hillier, 2009), Deleuze (Hillier, 2011; Wezemael, 2008) and Latour (Beauregard, 2015; Boelens, 2010; Rydin and Tate, 2016), as well as interesting work on complexity theory (Chettiparamb, 2014, 2019; de Roo, 2018) and institutionalism (Salet, 2018a, 2018b) are not discussed at length.[1] A growing number of anthologies (Fainstein and DeFilippis, 2016; Gualini, 2015; Madanipour, 2015), companions (Gunder et al., 2018; Healey and Hillier, 2016) and textbooks (Allmendinger, 2017; Beauregard, 2020) provide gateways to some of these new lines of thought should it be sought. It is not the function of this book to provide a similar survey of this ever-expanding corpus. Therefore, a degree of careful selection is required. As such, the following two chapters seek to trace those perspectives in the amorphous field of planning theory that in their traction at particular junctures or across time have had the greatest influence on the course of debate within the discipline.

In following Healey (2006: 336–337), planning theory is here understood as a 'planning perspective and purposive orientation' from which the world is apprehended. In keeping with the overall objectives of the book, each 'planning perspective' is reviewed in the context of 'how' the world is conceived, 'what' dimensions of this world are seen to matter and 'why' they are deemed to count. Issues of what is to be done by whom, how, when, where, why and with what consequences are discussed to complete the jigsaw of each theory where necessary. Examining this constellation of aspects facilitates working through the statements of leading proponents of each 'perspective' to determine and discuss its 'purposive orientation' by an examination of how each relates to a sense of the 'common good'. Achieving this involves discerning the suppositions underpinning each perspective to distinguish if, how and in what ways the concept of a common good sustains that theoretical proposition. This is undertaken to explore and highlight the relevance of the common good as an idea to planning theory, even in instances where this is only implicitly acknowledged, or even explicitly, rejected. Reconceiving this relationship of theory and practice to the common good is subsequently addressed in Part 2 of this book. However, as argued in Part 2, appreciating the role of the common good in planning requires sensitivity to the moral agency of the planning subject. Hence, in exploring the mixed terrain of planning theory, the following two chapters also examine the conceptual constitution of the planner that is detailed, implied or can be deduced from each perspective under review.

While it is acknowledged that structuring the nebulosity of planning theory is a largely artificial exercise that risks silencing the complexity of influences between perspectives, if done carefully it can nevertheless yield the benefit of balancing lucidity, breadth and depth in a succinct manner without compromising an accurate representation of the theories under review (for example, see Allmendinger, 2017; Beauregard, 2020; Gunder et al., 2018; Healey and Hillier, 2016). In keeping with this, the first of the two chapters in Part 1 examines those influential theoretical perspectives that share a tacit or overtly focused approach to the identification of the common good. Hence, Chapter 1 opens by examining attempts to render planning a scientific activity and how the response that this elicited spawned an influential field of theory focused on inclusive deliberation. Whereas at first glance these approaches seem wholly opposed, it is argued that they are threaded together by a supposition that the common good should be an issue of concern and is possible to identify. The theoretical perspectives reviewed in the subsequent chapter take an opposing position. These largely eschew the idea of a 'common' good, viewing it as at best a redundant concept and more frequently as one that is inimical to the expression of difference. Each of these perspectives has a particular concern with issues of subject formation, oppression and freedom. It is argued that the explanatory potential of each perspective reviewed in Chapter 2 is hamstrung by either its explicit suspicion of claims to the 'commonness' of any good or the difficulties in rendering its concept of the subject operable in a planning context.

So where to start? It can be argued that planning has a deep past stretching back to antiquity and encompassing any occasion in history where people have sought to spatially organise their interactions with each other and their world. However, 'planning theory' as an identifiable field is normally considered to have emerged following the Second World War, albeit respectful nods are usually given in books, papers and university courses to the 'inspirational precursors' (Hillier and Healey, 2008a) of 'architecture writ large' (Taylor, 1998: 159) such as Howard, Geddes and Le Corbusier, as well as to great administrative innovators, such as Selznick and Mannheim. Furthermore, this book focuses on planning in modern western democracies, which themselves are historically specific polities that have largely taken shape after the Second World War. Accordingly, the analytical narrative that follows begins in the 1950s, when the development requirements of a rubble-strewn Europe and booming North America created opportunities for the institutionalisation of new ideas on how planning ought to be done.

Note

1 While work using Latour and Deleuze is well developed in geography, such work in planning theory is growing but nevertheless still limited to a small number of authors.

1
A CONCEIVABLE COMMON GOOD

Planning as a Rational Scientific Activity

In an influential paper first published in 1985, Klosterman notes that academic and public attention to planning peaked during the 'great debate' of the 1930s and 1940s. He remarks that during this period defenders of free-market ideologies like Hayek and von Mises contended for dominance with the proponents of government planning, such as Mannheim and Wootton. Referencing Dahl and Lindblom (1953), Klosterman infers,

> By the 1950s the debate had apparently been resolved; the grand issues of the desirability and feasibility of planning had been replaced by more concrete questions concerning particular planning techniques and alternative institutional structures for achieving society's objectives. Planning's status in modern society seemed secure; the only remaining questions appeared to be 'who shall plan, for what purpose, in what conditions, and by what devices?'
>
> *(Klosterman, 1985: 5)*

For members of those affiliated with the Chicago School (University of Chicago), which is credited with seeding much thinking on planning following the Second World War, the answers to these questions were clear: planning was to be a technical discipline of societal management conducted for the common good that was characterised by coordinated government intervention and populated by experts (Sarbib, 1983). Taking their cue from fields such as economics and management sciences, champions of such 'big planning' like Rexford Tugwell believed that this expert-driven discipline would 'achieve a clear vision of the future above the din of petty politics, by becoming institutionalized as a fourth branch of government, with its own autonomous sphere' (Friedmann, 1987: 109). Conceived in this manner,

DOI: 10.4324/9781003155515-3

planning was to advance societal development through practices 'informed by scientific methods of inquiry and conducted in open and transparent ways' (Healey and Hillier, 2008: 299). This would redress issues of political manipulation too frequently evident in North American urban governance (Meyerson and Banfield, 1955). It is against this backdrop that positivist decision-making concepts emerging in allied professions were imported into planning during the 1960s and flourished in the early 1970s, marking what Taylor (1998: 71) identifies as 'the high tide of modern thought – the crest of that wave of optimism about the use of science and reason for human progress which had formed the European Enlightenment of the 18th century.' This reshaping of planning entailed a fundamentally different understanding of the practice than what had come before. As explained by Hall (2014: 295), 'Instead of the old master-plan or blueprint approach, the new concept was of planning as a *process* . . . And this planning process was independent of the thing that was planned.' To the forefront of such thinking were two distinct theories on how planning should be conceived, which as noted by Taylor (1998) are not, nor were not in the 1960s, always properly distinguished from each other. These are, namely, the 'systems' and 'rational' theoretical positions.

Systems Theory

The physically and aesthetically focused 'blueprint' approach to planning as 'architecture writ large' (Taylor, 1998: 159) came under increasing criticism throughout the 1960s due to its perceived inflexibility to an increasing awareness that cities where multifaceted and ever changing. Particular criticism was levelled at planning for its overriding concern with the 'appearance' of spaces and the scant regard shown to the mundane workings or experiences typifying these spaces. In this moment of reflection, prominent commentators such as Jane Jacobs appeared to expose the questionable theoretical suppositions upon which planning was founded, thereby casting doubt on the legitimacy of its professional expertise (Hirt and Zahm, 2012). It was in this context that a growing number of academics saw in the systems thinking of cybernetics a means to rectify these deficiencies. Indeed, as noted by Brian McLoughlin, one of the most prominent advocates of this new approach,

> In medicine and in management, in astronautics and in biology, cybernetics is showing its astonishing powers. It is able to do this because its field is the study of complex and probabilistic systems and their control.
> *(1969: 91)*

Inspired by new understandings of complexity in the emerging field of ecology and influenced by the enviable regard given the engineering-based mathematical models of transport planning, systems thinkers gained ground at a point in the history of modernist enthusiasm characterised by increasing computational power and an associated quantitative revolution sweeping across the social sciences (Giddens, 1990). For such theorists, the relevant dimensions of planning needed to extend beyond the physical and aesthetic to the measurable social and economic dynamics of urban life. This they felt would provide a more realistic perspective of how cities worked and as a consequence facilitate better planning (Forrester, 1969). Hence, behind the appeal to science, systems thinking was as much a normative endeavour as it was a project to furnish a more informed and intellectually robust approach to planning. In essence, it was believed that conceiving cities as 'systems' would enhance planners' capacity to effect change for the common good, thereby securing the position of the discipline as that which legitimately intervenes in the dynamics of urban change.

For systems theorists, urban environments exemplify a complex and fluid set of relationships, which much like ecosystems are nested within hierarchies of interaction. Here, the dynamics of one set of relationships can influence others to generate aggregate forces that alter the trajectory of activity in a city. To proponents of systems theorising, this new urban ontology represented a sea change in thinking that portrayed planning 'as an ongoing process of monitoring, analysing and intervening in fluid situations, rather than an exercise in producing "once-and-for-all" blueprints for the ideal future form of a town or city' (Taylor, 1998: 62). From this perspective, planning involves the production of mathematical models that 'make statements about the environment' (McLoughlin, 1969: 223), which facilitates the determination of desired development trajectories and the identification of appropriate interventions to realise these. Accordingly, the planner's role with respect to this reconceived city of dynamic relationships is transformed. As described by Ratcliffe,

> It is the planner's function to comprehend this tangled web of relationships, and where necessary, to guide, control and change their composition. To do this, planning is concerned with prediction, not only of population size and land use in isolation, but also of human and other activities as well. It has been said that planners are now the prisoners of the discovery that in a city everything affects everything else.
>
> *(1981: 115)*

In this sense, 'the systems perspective is highly "planner-centric" in that it places a great deal of emphasis upon professional opinion in an abstracted and technical process where goals flow from an analysis of problems' (Allmendinger, 2017: 64). Planning and planners are located like the brain of a system to which all information flows, is processed, and direction subsequently issues. With a focus on quantitative data, the information collated is used to model possible trajectories and action then taken to calibrate decisions such that plans are conceived as 'charts of a course to be steered' (McLoughlin, 1969: 83). This newfound expertise placed planners in an authoritative position of direction-setting justified on the view that

> one of the most forceful arguments for placing primary responsibility for goal formulation on the planner . . . [is] the assumption, traditional to profession-

als, that, in some way, they 'know more' about the situations on which they
advise than do their clients.

(Chadwick, 1971: 121)

No longer bound by quickly outdated 'blueprints,' planning was reimagined as an
open-ended endeavour of 'piloting' wherein the planner both sets the course and
corrects for changes along the way. As summarised by McLoughlin,

> We can picture the planner now as a helmsman steering the city. His (sic)
> attention focuses on the plan – the charted course – the future states through
> which the city should pass – and on the survey observations which indicate
> its actual state.
>
> *(1969: 86–87)*

The planner thus tames the fluid complexity of urban dynamics via the power
of analysis. Yet, by both charting the course and piloting, this approach smuggles
into the seeming objectivity of the process a degree of intersubjective normativity wherein expert decisions on what should be counted, how, when and why
ultimately leads to decisions on what amounts to the common good ('the charted
course'), and how this should be realised by 'steering the city'. Hence, it is unsurprising that this perspective was criticised for the hubristic assumption that it was
possible for planners to comprehensively quantify, understand and control the fluid
complexity of cities in determining and delivering the common good. Most notable among this reproach was the criticism that extracting planning from the clutter
of politics silenced the messiness of real-world urban governance where frequently
unpredictable and conflicting imperatives rendered the objectives of a systems perspective unachievable. For some of its contemporary critics, the problem of systems
theory thereby lay not in the assumption that a 'scientific' approach to planning was
at fault. Instead, the primary issue was how the scientific method had been misconceived by systems perspectives to incorporate goal specification. Accordingly,
another camp of theorists stressed the need to confine the ambitions of planning to
the *process* of effectively delivering on political decisions. This view is most generally known as the 'rational process theory' of planning.

Rational Process Theory

Prominent among theorists who believed that planning should be instrumental in
serving the implementation of goals rather than in specifying them was Andreas
Fauldi (1973). His influential work became a touchstone in planning thinking from
the mid-1970s well into the 1980s. Fauldi was enamoured by the scientific method,
which he equated with planning. As he exclaimed,

> I perceive planning as analogous to that other activity that has resulted in
> unparalleled human growth: science. Planning and science can be seen as

twin sisters born from the same desire of man (sic) to free himself from the strictures of ignorance and fear. Planning and science propel this process of man becoming master over his world and himself along a path towards further human growth.

(2008 (1973): 375)

In focusing on science as a 'method' for objective knowledge production rather than means for determining what ought to be done, Faludi urged planners to envisage planning as an instrumental activity grounded in 'process'. The concern here was the irrationality characterising politics that made prediction uncertain at best and often impossible. Thus, following Max Weber's appeal that formal and substantive rationality should be kept distinct (Kalberg, 1980), Faludi sought to conceive a type of planning that would realise what Mannheim (1940: 267) termed 'the rational mastery of the irrational' by partitioning ends (substantive rationality) from means (formal rationality) in creating an impartial practice for establishing the most efficient way to realise goals. In contrast to systems theory, the goals themselves were not to be set through comprehensive evaluation 'in' planning, although analysis could inform goals. Instead, goals were established in advance of the application of rational planning through democratic political processes that then 'fed into' planning (Dror, 1968). Planning was thereby reimagined as a 'set of procedures' (Davidoff and Reiner, 1973: 11) that are selected as the best way of delivering on goals (Faludi, 1973: 5). As summarised by Healey and Hillier (2008: 300),

> Starting from goals to be achieved, impediments to goal-achievement are analysed using 'scientific' methods. From this analysis a range of policy options can be identified. These are then evaluated by criteria derived from value premises embodied in the goals. Such a linear and logical form of argumentation gives transparency to decision-makers and those who judge them, and in theory, provides a countervailing force to political manipulation.

Resonant with Popper's philosophy of science (Popper, 1959), planning was thus conceived as a scientific enterprise concerned with the identification of alternative options (hypotheses) for the realisation of goals. The predictive testing of these options would determine which option is the most effective and efficient in delivering such goals. Carefully monitoring the implementation of the identified best option would subsequently facilitate feedback that could be used to establish where corrections are needed to efficaciously resolve impediments to goal attainment. Accordingly, in mapping across the ontology of science onto planning, issues of value, politics and personality were bracketed from practice such that 'planners should act much like research scientists in searching for the best methodology' (Allmendinger, 2017: 70). From this perspective, the planning subject is viewed as a neutral scientist whose impartial reasoning can be used as an instrument for delivering the common good. However, the conception of what this common good might be lies beyond the scientific objectivity of rational planning and is instead determined through processes of democratic politics. As such, the normativity of planning is bounded by the model of how planning ought to be done rather than what substantive ends planning ought to achieve (Faludi, 1973: 116).

By reconceiving planning as a rational scientific activity, both the systems theory and rational process perspectives reflected a fundamental change in how planning should be envisaged. Both perspectives share an assumption of the ontology of an objective world wherein a predictable constellation of attributes and forces can be identified and most effectively addressed through the proper application of scientific methods. As summarised by Taylor (1998: 70),

> One way of describing the change in planning thought which occurred from the 1950s to the 1960s is that, whereas in the 1950s and before, town planning was seen as primarily an art, by the end of the 1960s it had come to be seen as primarily a science.

Although sharing a focus on the importance of scientific procedure, the two theories nonetheless differ regarding the position of substantive rationality, with systems theorists comfortable in specifying the common good, while those of the rational process school confining planning to the realisation of decisions taken elsewhere on what the common good may entail. That said, for proponents of rational process, advancing better process is inherently entwined with 'human growth' as the implicit common good underpinning the project. As explained by Faludi,

> Growth as a *process* refers mainly to learning and creativity, defined as the gaining of insights into the existing order of things, and the transformation of that order into a new one. . . . It is that process by which man (sic) creates himself which brings us to reasons for putting forward human growth as the rationale for planning theory.
>
> *(2008 (1973): 381)*

Thus, rather than a linear trajectory from the natural sciences to planning theory, supporters of planning as a rational scientific activity 'struggled over the nature of reason and the complex relations between ends and means, facts and values' (Healey and Hillier, 2008: 304). Yet both the systems theory and rational process perspectives emerging from this venture posited 'a high degree of control over the decision-making situation on the part of the decision-maker' (Etzioni, 1967: 385). This led some to question the appropriateness of such an accumulation of power. These critics noted how the ostensible scientific detachment of such perspectives nevertheless involved various assumptions made by so-called experts during the decision train. As described by Healey and Hillier (2008: 303),

> throughout the process of apparent logical deduction from goals to analyses, the formulation and evaluation of means and the structuring of choices for politicians, all kinds of assumptions and logical leaps had to be made by

the 'experts'. How these were made depended significantly on conceptions and values locked into particular epistemologies used to describe systems and analyse relations.

If values are 'locked into' ways of describing and analysing, it follows that the 'facts' of planning are not so much impartial as intersubjective. Consequently, it was argued by some that 'Appropriate planning action cannot be prescribed from a position of value neutrality, for prescriptions are based on desired objectives' (Davidoff, 1965: 331). The implication of this view is that if facts and values are interwoven, different interpretive frameworks contour what is valued and why. Accordingly, planning to promote 'human growth' should not be sought via the 'objectivity' of technical expertise, as this would inevitably be framed by particular worldviews that are blind to the multitude of non-scientific ways the world is interpretated. The conclusion was thereby drawn that better planning is more about enhancing the plurality of democratic input to decision-making than it is about formulating more accurate scientific methods. Underpinning much reflection on this in planning theory has been pragmatic philosophy.

Deliberation and Diversity

Pragmatism

Pragmatism is a philosophy that is suspicious of grand explanations offered by approaches that advance an all-encompassing account of why things are the way they are. It is dubious about deductive forms of reasoning that place principle above the activities of thinking and doing which influence peoples' engagement with knowing, deciding and acting in the context of lived experience. Indeed, Hoch (2018: 119) notes, 'Instead of trying to redeem the salience of philosophy as the intellectual foundation for scientific reason, the pragmatists embraced the practice of scientific inquiry as the focus for philosophical reflection.' This philosophical approach was developed in North America and is most associated with three thinkers. As summarised by Thayer (1981: 5),

> In a word, pragmatism is a method of philosophizing often identified as a theory of meaning first stated by Charles Pierce in the 1870's; revived primarily as a theory of truth in 1898 by William James; and further developed, expanded, and disseminated by John Dewey.

Pierce in particular was fascinated by the emerging 'scientific method' of the late nineteenth century. For him, the means of philosophising characteristic of contemporary academia was misplaced in its foundationalism and level of abstraction. He advanced a humbler approach which held that our understanding of the world was constituted through engagement with it such that understanding is instrumental in providing knowledge on what is true. In this sense, what matters as fact is conceived *a posteriori* to the lived life rather than *a priori* via rational detachment. Truths are products of the consequences they offer in helping us negotiate a world of knowledge requirements. Yet such truths are provisional and subject to correction. Pierce encapsulates this view in his pragmatic maxim:

> Consider what effects, that might conceivably have practical bearings, we conceive the object of our conception to have. Then our conception of these effects is the whole of our conception of the object.
>
> *(Peirce, 1955: 31)*

James reinterpreted and extended Pierce's concept to rework the pragmatic maxim in a phycological direction. For James, a truth is delineated by its effectiveness in resolving doubt. Accordingly, whereas for Pierce the veracity of a proposition is determined through experience, for James the truth of a proposition is defined by the implications of adopting or believing in it. To some extent, this shift in positions can be explained by James' interest in religions and the emerging field of psychology. Nevertheless, its effect was to extend pragmatism into the social world where the intersubjectivity of facts and values are more to the fore of debate. In this sense, it recentred much thinking in pragmatism from a focus on describing the truths of material existence to a concern with norms, relationships and the politics of knowledge. Dewey in particular was concerned with how understanding was produced through interaction, claiming that 'the object of knowledge is eventual; that is, it is the outcome of directed experimental operations, instead of something in sufficient existence before the act of knowing' (Dewey, 2008: 136). Hence, these early pragmatic thinkers promoted what would today be termed a social constructionist ontology (Berger and Luckmann, 1966) that emphasises 'a relational view of the world, without losing a deep understanding of its material solidity' (Healey, 2009: 281). However, Dewey developed this beyond epistemological reflection through a political turn on the works of his forebearers. He was specifically concerned with how the processes of participatory democracy when refracted through a pragmatic understanding of knowledge could enhance deliberative practice as a force for good. With a strong humanist bent, Dewey inspired theorists to seek solutions in the agency of actors rather than in the dupes of Marxist dogma and the ossifying abyss of structuralist thinking. He did this by stressing the importance of attention to how judgements are made in real-world contexts that combine analytical, emotional and moral reasoning. Here, context was to be implicated in the decisions reached through intersubjectively seeking clarity in an ever-emerging horizon of issues to be negotiated. Knowledge and action produced by a community of knowers are what matters; enhancing the relationships inherent to facilitating this is what counts. As social relations and polities were placed centre stage, the role of knowledge production was de-centred from the preserve of experts to that which is communally produced in context. It was this concern with how knowledge is produced and used that began to influence planning in the late 1960s and early 1970s.

While not ostensibly associated with any pragmatic thinkers, Lindblom's (1959) 'branching' incrementalism and Etzioni's (1967) 'mixed-scanning' are sometimes referenced as early examples of pragmatic thought in public administration. However, it was not till John Friedmann's combination of institutional thinking with Deweyan perspectives on democratic potential that the benefits of a pragmatic approach for planning were coherently adumbrated. Friedmann developed his views when reflecting upon his work during the 1960s in advising Latin American governments on regional and urban strategies. He came to the conclusion that the model of planning as a scientific activity was neither a reflection of what happens in practice nor what ought to occur. His experiences made him realise that under the scientific model 'planners talk primarily to other planners, and their counsel falls on unresponsive ears'. Ultimately, he concluded,

> The establishment of a more satisfactory form of communication is not simply a matter of translating the abstract and highly symbolic language of planning into the simpler and more experience-related vocabulary of the client. The real solution involves a restructuring of the basic relationship between planner and client.
>
> *(Friedmann, 1973: 172)*

From this insight, he advanced a new approach that he called 'transactive planning'. This focused on a 'life of dialogue' as the means to cultivate a more effective means of understanding how planning could and should be constituted. Here, enhancing dialogue between planners and their publics was implicitly perceived as a common good that would promote a form of decision-making which is more responsive to contextually contingent conditions than detached 'scientific' models. For Friedmann, this would require that planners acquire new competencies. Accordingly, he asserted that successful planning 'depends in large measure on the planner's skill in managing interpersonal relations' such that

> the qualities he (sic) would have to develop include a heightened knowledge of the self; and increased capacity for learning; . . . a heightened empathy; an ability to live with conflict; and an understanding of the dynamics of power and the art of getting things done.
>
> *(1973: 20)*

In this sense, Friedmann envisaged the transformation of planners from modellers of collated scientific data to flexible collaborators who plan with the public. The dialogue of mutual learning was to replace the monologue of expert specification.[1]

Meanwhile, Judith Innes had published her doctoral research into how the social indictors used in policy were produced through interactive processes of interpretation. She revealed how such indicators were often formulated in the context of technical uncertainties and political debate prior to their presentation as ostensibly objective measures (De Neufville and Innes, 1975). By unveiling the intersubjectivity and political construction of knowledge, her work thereby undermined the legitimacy of presenting planning as a scientific activity profiled by impartial technical expertise. More focused on the individual planner than governance processes, Donald Schön was concurrently reflecting on how professionals could and should use their expertise. Influenced by Dewey (Blanco, 1994), Schön advised practitioners to explore how experiences of trial-and-error influence decision-making in public policy (Argyris and Schön, 1974; Schön, 1971). He developed this vein of thinking into a book-length consideration of the creativity that arises through 'knowing-in-action' when professionals negotiate challenges in their practice (Schön, 1991). In subsequent work with Martin Rein, Schön emphasised the importance of conceiving policy practice as

> a kind of distributed designing, undertaken by multiple actors in the policy environment, and to think of policy rationality in terms not only of rational choice, or even rational politics or negotiations, but of a more encompassing kind of rationality, inclusive of higher-level reflection.
>
> *(1994: xviii)*

Through this process of reasoning from multiple perspectives with many actors, the frames anchoring intractable policy problems could be transformed, thereby facilitating resolution. To enable the full appreciation of what can be learnt from different perspectives, some researchers working with pragmatic ideas began focusing on the roles played by power and inclusion/exclusion in the formulation of knowledge, agenda setting and the trajectory of decision-making (Hoch, 1984). Influential theorists such as Hoch (1996: 42) suggested,

> Instead of relying on rational methods that require practitioners to seek greater political authority and more professional power in order to do good, planners might benefit more from a critical review of the limits to bureaucratic command and adversarial democracy. Planners might consider identifying with the powers of the weak, identifying with colleagues, neighbours, and citizens rather than with the protocols of professional expertise.

Foremost among those thinking on these lines is John Forester. He combines a pragmatic approach with the critical theory of Jürgen Habermas to formulate a power-sensitive perspective he terms 'critical pragmatism'.

Critical Pragmatism

Forester studied at UC Berkeley in the 1970s where he was involved in various progressive movements influenced by the radicalism of the late 1960s. During this period, he developed misgivings about systems thinking that portrayed planning as a scientific activity. For him, this form of rationality failed to account for what he was experiencing through his various interactions with voluntary organisations (Forester, 2017). It was against this backdrop that he became interested in the post-positivist philosophy of Wittgenstein (1953), the pragmatism of Dewey and the critical theory of Habermas (1971, 1979). He combined insights gleamed from these and other theorists in language philosophy (Austin, 1962), political theory (Pitkin, 1972) and pragmatism (Bernstein, 1983) 'to demonstrate how planning conduct involves much more than judgements based on rational method and principles' and to show how 'planning practitioners perform and conceive advice within linguistic, social and cultural deliberation' (Hillier and Healey, 2008b: 4). Forester is concerned with highlighting the inherently political but transformative potential of how understandings are produced through the micro-dynamics of people interacting in specific contexts. Following Dewey, he believes that the aggregative impact of changed perceptions and practices at the small scale can result in overall improvements to how planning operates more generally. However, as noted by Healey (2012a: 342), it is important to bear in mind that for Forester, planning

> is about mobilizing 'attention' and about generating 'hope' that it is possible, by some form of intervention directed at public purpose, to improve conditions . . . . It is a project that involves combining technical capacity with moral purpose and sociopolitical sensibility.

Forester thereby views planning as an effort to advance open and free deliberation that can stimulate a shared clarity and optimism for change in the face of those challenges posed by 'political inequality, racism, turf wars, and the systematic marginalization and exclusion of the poor' (Forester, 1999a: ix). Building on the base of a social constructionism inherited from pragmatism, he harnessed Habermas' critical theory as both an ontological perspective and guide to address the problem of power asymmetries, which he believes hinder the realisation of planning's democratic potential.

In broad terms, Habermas sought to remake the 'unfinished project of modernity' (d'Entrèves and Benhabib, 1997) in a humanist vein by interrogating the everyday dominance of instrumental rationality in contemporary society and accentuating alternative means for thinking and knowing. He reasoned three forms of rationality: technical-instrumental, emotive-aesthetic and ethical-moral. Habermas traced how the technical-instrumental form of rationality had come to 'colonise the lifeworld' – the 'lifeworld' being the term he used to describe the world experienced phenomenologically through our relations with others in sharing knowledge and coordinating social activities. He argued that this 'colonisation' resulted in restricted intersubjective understanding, recognition and respect as the language and rationalities of technical instrumentalism came to profile thinking and action in ways that produce inequities of representation. Forester borrowed from Habermas the concept of 'ideal speech,' which Habermas constructs on the insight that all dialogue is characterised by a latent truth. As Habermas asserts,

> No matter how the intersubjectivity of mutual understanding may be deformed, the design of an ideal speech is necessarily implied in the structure of potential speech, since all speech, even of intentional deception, is orientated towards the idea of truth.
>
> *(1970: 372)*

Habermas does not hold fast to the notion that 'ideal speech' is easy or even possible to achieve. Rather, he aspires that an enlightened modernity can progress towards this ideal, and in doing so temper distortions of power in representation. He thereby construes the ideal speech situation as one that

> excludes all force – whether it arises from within the process of reaching understanding itself or influences it from the outside – except the force of the better argument (and thus that it also excludes, on their part, all motives except that of a co-operative search for the truth).
>
> *(Habermas, 1985: 25)*

Habermas builds on this understanding to establish the normative prerequisites for an 'ideal speech situation' – namely, truthfulness, sincerity, legitimacy and comprehensibility (Habermas, 1987; Low, 1991a). Hence, Forester 'sets out a research agenda to test how recognition of these prerequisites (and their abuse) might provide a resource for comprehending and improving planning advice in different contexts' (Hillier and Healey, 2008b: 5). In doing so, he ventures to 'redefine the role of the planner away from a handmaiden to instrumentalist, rational decision makers to [a] hands-on professional who fosters inclusive, participatory forms of collective action' (Wagenaar, 2011: 293). In essence, Forester is arguing for more open and democratic processes where the planner plays an active role in exposing and challenging powerful interests that marginalise different viewpoints (Allmendinger, 2017: 138). As he asserts,

> The planner's responsibility to address possibilities of legitimate public policy calls then for work redressing or circumventing unnecessary structural distortions of communications: deliberate exclusion of affected publics, repression of available information concerning policy consequences, ideological justifications of policies, incomprehensible or obscure bureaucratic language, and so on.
>
> *(Forester, 1993: 28)*

With a profound commitment to 'cultivating a progressive democratic polity,' Forester 'demands a particular value orientation from planners' (Healey, 2012a: 342) such that 'the talk and listening of planners is at once practical, interpretive and deeply political' (Forester, 1989: 137). It is this sensitivity to power and his desire to attend to it that gives Forester's approach its 'critical' edge. As he outlines,

> by treating planning as inherently political and communicative, a critical theory of planning practice can: (1) be grounded in an empirical analysis of what

planners do intentionally and unintentionally; (2) be sensitive to the practical situations which planners face and seek to interpret and understand; and (3) be discriminatingly critical of the extent to which planners counter (or perpetuate) unnecessary political distortions of problem formulations, analyses of options, or broader planning agendas.

(Forester, 1993: 16)

Thus, by reconceiving planning and the role of the planner in this way, Forester moves his critical approach towards a concept of the common good that entails equality of participation in open fora of deliberation grounded in a liberal plural democracy which fosters consent in the resolution of complex and sometimes contentious planning decisions. Doing good planning thereby requires that planners work to develop a politics of 'true consent,' wherein 'such consent grows out of uncoerced collective criticism, political argument, and dialogue' (Forester, 1989: 141).[2]

Other theorists to emerge from Berkeley during the 1980s likewise advanced a pragmatist approach to planning, although their work is less focused on remedying 'systemic distortions of power' than Forester's 'critical' approach. For example, Hilda Blanco drew on Pierce's concept of 'abduction' to demonstrate how planning activity begins when a problematic issue is encountered in a specific context, rather than with abstract goals. Employing this pragmatic sensitivity to conceive planning as a form of situated problem solving, the role of the planner is understood as facilitating deliberation in devising and considering alternative options for action. With reference to Dewey, she emphasises democratic inquiry as a form of imaginative planning aimed at 'developing a language that could reanimate a meaningful public realm' (Blanco, 1994: 164). Accordingly, for Blanco, the inferred common good is the condition of open democratic interrogation where the intangibility of 'vague' and 'general' public goals are given specificity through their translation and instantiation into situations that people can identify with and properly debate.

An important theorist to also emerge from Berkeley during this period is Niraj Verma. His work traces an eclipse of the pragmatic perspective by analytical philosophy during the mid-twentieth century. He claims that this gave impetus to conceiving planning as a scientific activity resonant with systems thinking and the rational process approach (Verma, 1996). Verma's review provides a platform from which to argue that pragmatism is more apt at tackling the multitude of 'wicked problems' (Rittel and Webber, 1973) populating planning than is the formal logic of analytical philosophy. However, instead of mining the work of Dewey as a beacon for thought, Verma expounds the value of James' work in helping us reconceive how problems are framed and solutions determined. This involves close attention to undoing the fact-value dichotomy that had emerged in planning consequent on the dominance of analytical philosophy. In particular, Verma draws upon James' interest in the phycological dimensions of epistemology. He notes how this influences the acceptability of what is deemed a 'truth', such that facts about experiences can be understood more as the readiness of an audience to accept them than their objective veracity. Hence, Verma notes how, 'Truth is not only a matter of logical agreement between different elements of knowledge; it is also the psychological agreement between messenger, the message, and its receiver' (1996: 6). What interests Verma is thus a focus on how meaning is made and circulated. Informed by James, he contends that a 'truth cannot be understood outside of the sociological and psychological processes' in which it is produced and disseminated, such that a 'knowledge of community affects our knowledge of truth and finding out what is the truth affects our knowledge of community' (1996: 10). He follows James in suggesting that we call assumptions into question by locating phenomena comparable to what we are confronted with and making analogies that offer new vistas on how to conceive planning as an experience of tackling challenges. This enables the revision of assumptions impeding the resolution of those issues planning is faced with (Verma, 1993, 1998). Accordingly, the role of the planner is to facilitate the exploration of analogies and alternatives for reworking our understandings of the matters at hand. Although a common good is not clearly discernible from this Jamesian-informed pragmatic understanding of planning, it is nevertheless implied as procedural rather than substantive and focused on a process that is contoured by a commitment to the resolution of persistent complex problems through finding ways to think issues anew.

In Canada, Tom Harper and Stan Stein had been working separately but in parallel to develop a view of planning inspired by neo-pragmatic ideas (Davidson, 1984; Putnam, 1981; Quine, 1969; Rorty, 1991), yet most intensely by the influential writings of John Rawls (1971). Rawls' focus on the theory of 'justice as fairness' helped shape the concepts they developed on planning. In reconceiving what 'justice as fairness' might mean, Rawls advances a cogent yet deceptively simple argument that seeks to locate principles 'to govern the assignment of rights and duties and to regulate the distribution of social and economic advantages' (Rawls, 1971: 61). His strategy in pursuing these principles is to formulate a thought experiment that imagines an 'original position' and 'veil of ignorance' whereby ensuring fairness is connected to ignorance, such that people are asked to develop the principles of distribution governing a world into which they will enter not knowing the circumstances into which they will be born (rich, poor, gifted, average, etc.). In Rawls' view, people in the original position, denied knowledge of their aptitudes and endowments, will not be driven by a certain concept of 'the good' but instead by interest in devising a set of principles that guarantees their liberty, greatest socio-economic benefit and a capacity to attain positions of wealth and influence. For Rawls, this will logically result in people choosing to formulate a series of rational principles that maximise fairness by instituting a system of justice which supplies the most extensive organisation of basic liberties for everybody and furnishes equality of opportunity for all (Rawls, 1971: 302). Of note is that in Rawls' view the concept of the 'good' is to be determined subsequent to the establishment and institutionalisation of these principles. Hence, 'the good' is to be 'produced' following the institutionalisation of the principles of justice through debate among people enjoying equality of opportunity, rather than pre-emptively steering deliberation on what principles of justice should be formulated and institutionalised (Lennon, 2017; Sandel, 1982). Through this line of reasoning,

> Rawls has been so influential because, within a vocabulary acceptable to proponents of rational choice theory, he presents a logical argument that defends equality of primary goods as the basis of justice without resorting to natural law, theology, altruism, Marxist teleology, or a diagnosis of human nature.
>
> *(Fainstein, 2010: 15)*

Harper and Stein adopt this logic to argue that planning theory should be constructed on the ontological assumption of 'the free, equal, and autonomous individual person as the basic unit of society – the ultimate object of moral concern and the ultimate source of value' (1996: 12). They hold that 'an appeal to the autonomous individual is necessary for critical rationality' (p. 13). Advancing such 'political liberalism' as the basis for planning theory (Harper and Stein, 2006), they seek to partition the 'private realm' (personal beliefs) from the 'public realm,' as it is in the latter that they 'are concerned not with whether an idea is true or false (arriving at the 'truth of the matter') but with whether the idea can command consensual support as providing a reasonable basis for public policy in a democratic society' (Stein and Harper, 2005: 150). Their project is to build upon this separation an approach that is 'liberal in its normative values, critical in its assessment of current planning practices, and incremental in its justification and practice' (Healey, 2009: 286). The upshot of their thinking is the moulding of a planning subject capable of detachment from their private views in performing professional duties. As they note, 'It is because of the internal relationship between professional planning and its moral goals that professional planners should have a value-neutral role' (Stein and Harper, 2005: 151). Thus, in contrast to the approach adopted by other theorists informed by pragmatic philosophies, Harper and Stein retreat from a focus on practice to theorise a series of *a priori* principles that both amount to and help realise something roughly equating to the common good. Consequently, their approach has been criticised by some of those working with pragmatic planning theory as more ivory tower idealism than pragmatic engagement (Alexander, 2008; Hoch, 1993). Indeed, while the work of Harper and Stein is frequently cited in writings on planning theory, few have sought to mobilise it in the interpretation of empirical material (however see McKay et al., 2012).

Across the Atlantic in Europe, planning theorists were also beginning to mine the works of pragmatic thinkers. For example, research by Tøre Sager into transport planning controversies in Norway drew on Lindblom's concept of 'disjointed incrementalism' (Lindblom, 1959), Friedmann's 'transactive planning' (Friedmann, 1973) and Forester's 'critical pragmatism' (Forester, 1989, 1993) in the formulation of his 'dialogical and communicative rationality' (Sager, 1994: 20). Mobilising Habermas, he sought a means to square technical expertise with open deliberation in ways that promote working with and through the realities of imperfect knowledge in politically charged situations. Here it is the job of the planner to facilitate the reconciliation of goal-driven instrumental and communicative rationalities so that both technical and lay knowledges can be successfully brought into planning (Healey, 2012a: 343). As such, Sager pursues a reconceptualisation of how planning can operate. By treating the rationalities of science and deliberation as equals in a process of open knowledge production and solution identification, he implies a common good, which much like that of Forester, is tacitly presented as the production and sustenance of fora for consensus-seeking debate based on recognition, reciprocity and a respect for difference.[3]

Hence, by the 1990s, pragmatism was a major force in planning theory, especially in North America. As an alternative to systems thinking, rational proceduralism and structuralist explanations of society, it offered a way out of the seemingly intractable disputes that dogged practice. By the close of the twentieth century, it had been extensively mined to help develop a theoretical programme that was sophisticated in its ontological grounding, mature in its conception of planning and the role of the planner, and progressive in its understanding of what the common good may entail, albeit this was often implicit rather than explicitly stated. Most notably, the influence of pragmatism on planning theory is characterised by the view that the politics of disagreement is inherent to planning and so must be dealt with constructively. As summarised by Hoch (2019: 2),

> The pragmatist approach does not avoid moral and political conflicts, differences in social standing or asymmetries of political power. It treats these as inescapable conditions like physical features of the environment, demographic trends, transportation technology and the multitude of influences shaping what the future holds. The pragmatist outlook looks to the future, imagining horizons for social improvement that rely upon intelligent democratic collaboration.

With democratic collaboration placed centre stage, 'the big question for the pragmatic analysts is how practitioners construct the free spaces in which democratic planning can be institutionalized' (Hoch, 1996: 42). Patsy Healey was to address this question by exploring avenues for the institutionalisation of deliberative approaches in planning activities through her development of what became known as 'collaborative planning'.

Collaborative Planning

At first glance, critical pragmatism and collaborative planning may appear as synonyms. Indeed, as with critical pragmatism, 'The collaborative planning idea promotes the significance of careful attention to the social and communicative relations through which any planning work is done and could be done' (Healey, 2012b: 75). Moreover, both draw upon Habermas and his normative endeavour to reconstruct the public realm in ways that impede domination of the lifeworld by technical-instrumental rationality. However, reflecting its development in the

United Kingdom, collaborative planning draws on social institutional theory rather than on the pragmatic philosophers who had influenced Hoch, Forester and others working in a North American context. Furthermore, collaborative planning offers a concerted attempt to address issues around the dynamics of urban regions that are sometimes lacking in critical pragmatism (Healey, 1996). This adds a greater spatial sensitivity to the politics of deliberation such that issues of 'place' are perceived to emerge as the product of frequently competing definitions that simultaneously support various meanings.

Echoing Friedmann's experiences, Healey's exposure to planning practices in Latin America, coupled with her background in anthropology, prompted her to adopt an interpretive approach in her research. She mobilised ethnographic methods in studying how planners manage the complexities of the tasks they were confronted with (Healey, 1992a, 2017). Building upon a series of papers written throughout the late 1980s and early 1990s, she brought her thinking together in a book-length statement titled *Collaborative Planning* (Healey, 1997). As she subsequently reflects,

> Conceptually, the book offers a social-constructivist and relational approach to urban and regional dynamics and governance processes. This was informed by a recognition of the multiplicity of social worlds, 'rationalities' and practices that coexist in urban contexts and the complexity of the power relations within and between them.
>
> *(Healey, 2003: 107)*

From this ontological perspective, people are understood to 'live in complex webs of social relations with others, through which cultural resources – ways of thinking, ways of organising and ways of conducting life – are developed, maintained, transformed and reproduced' (Healey, 2006: 44). The challenge Healey confronts is 'how to reconcile the individuation of cultural identity with recognition of commonality between individuals with different frames of reference, as well as different interests' in ways that 'helps to open up opportunities for diversity'. For Healey, as with those critical pragmatists operating in North America, the objective is to 'reconstruct a public realm within which we can debate and manage our collective concerns in as inclusive a way as possible' (Healey, 2006: 44). Her work is thereby highly normative in seeking to investigate

> the conditions under which particular forms of collaborative process may have the potential to be transformative, to change the practices, cultures and outcomes of 'place governance', and, in particular, to explore how, through attention to process design, such processes could be made more socially just, and, in the context of the multiplicity of urban social worlds, more socially inclusive.
>
> *(Healey, 2003: 107–108)*

A key innovation of Healey's approach is the utilisation of neo-institutional theory (Powell and DiMaggio, 1991; Scott, 2008), particularly the use made of Giddens' 'structuration theory' (Giddens, 1984). Drawing from Giddens, Healey contends that 'our individual identities and social relations are 'structured' by what has gone before,' which serve as 'active forces filled with implicit and explicit principles about how things should be done and who should get what' (Healey, 2006: 45). Healey believes that an important benefit of structuration theory is the recursive relationship between structure and agency. Accordingly, as we plan, 'we affirm our pasts, challenge them and change them' such that 'structures are "shaped" by agency, just as they in turn "shape" agency' (p. 46). It is in this sense that Giddens' theory of structuration 'emphasises that individuals are neither fully autonomous nor automatons' (p. 49). Healey identifies in Giddens' work a recognition that 'we "have power", and, if sufficiently aware of the structuring constraints bearing on us, can work to make changes by changing the rules, changing the flow of resources, and, most significantly, by changing the way we think about things' (Healey, 2003: 49). Thus, Giddens supplies the conceptual apparatus for the reinvestment of agency into a structure that is resonant with Healey's social-constructivist and relational ontological approach (Healey, 1999). With Giddens as a platform for describing 'how things are,' she harnesses Habermas' theory of communicative action to provide direction on 'how and why things ought to be'. As she reasons,

> the contributions of Giddens and Habermas, the one emphasising active agency in the power of structures, and the other focusing attention on the processes of collective dialogue and how to confront the distortion of dialogue by the powerful, highlight both the cultural boundedness of ways of thinking and acting, and the possibilities of learning, for development, and for transformative action.
>
> *(Healey, 2006: 54)*

From this theoretical context, Healey holds that,

> a progressive, normative meaning could be given to 'collaborative planning', as planning activity centred on working interactively with stakeholders with diverse stakes and on place development problems and futures in ways that recognise and respect multiple perspectives and modes of engaging in governance work and that promote inclusive and richly informed public policy making.
>
> *(Healey, 2012b: 61)*

Hence, the identification and inclusion of stakeholders are central to the success of the collaborative planning approach (Healey, 1998). Indeed, Healey believes that unless stakeholder contribution is fully acknowledged in the processes of planning, 'policies and practices will be challenged, undermined and ignored' (Healey, 2006: 70). However, for Habermasian-informed collaborative planning to work and endure, participants in planning activities 'must learn how to build consensus across their differences' (p. 70). By creating, consolidating and expanding the webs of relations that bind communities, it is contended that collaborative planning 'has a role in building up the institutional capacity of place' (p. 61). This progressive focus on agency in collaborative planning views institutional change in planning as emerging 'from the "grass-roots" of the real concerns of specific stakeholders as these interact with each other in specific situations in place and time' that for Healey 'produces institutional infrastructure which is as near as possible to the lifeworlds of the stakeholders' (p. 285).[4] Consequently, despite rejecting the view of 'planning as a scientific activity,' collaborative planning places procedure front and centre (Healey, 1992b). Indeed, Healey asserts,

> Spatial planning efforts should therefore be judged by the qualities of *process*, whether they build up relations between stakeholders in urban region space (sic), and whether the relations enable trust and understanding to flow among the stakeholders and generate sufficient support for policies and strategies to enable these to be relevant to the material opportunities available and the cultural values of those involved, and have the capacity to endure over time.
> *(2006: 71 – emphasis in original)*

From a normative perspective, the issue then becomes locating 'ways to evaluate the quality of the communicative and collaborative dynamics through which social relations are maintained and changed' (Healey, 2003: 112). Healey believes that this change in how planning activity is evaluated presents ethical challenges to planners with respect to what they know, how their knowledge is being used and the ways they conduct themselves. In essence, it requires 'a renewal of the expertise of those locked into the ways of thinking and acting of previous government practices.' Following this line of argument, she specifies,

> In this context, the traditional spatial planner is in many cases being transformed into a kind of knowledge mediator and broker, using an understanding of the dynamics of the governance situation to draw in knowledge resources and work out how to make them available in a digestible fashion to the dialogical processes of policy development.
> *(Healey, 2006: 309)*

However, such a reconceptualisation does not presuppose the continuation of the planner's centrality in planning activities. Indeed, 'it follows that planners adopt less central roles' (Harris, 2002: 40). This reimaging of planning thereby brings the conception of the planner a considerable distance from that envisaged in the 'systems theory' wherein scientific expertise was valorised or in the 'rational process' approach in which technical proficiency was lauded. Likewise, what constitutes the common good is also transformed. No longer viewed as simply some 'good' to be delivered *by* planning, the common good becomes the very nature *of* planning itself. In this sense, 'Collaborative planning practices are advocated not just for their instrumental value in making governance interventions in place development more effective, but also for their contribution to developing an inclusive polity in which the concerns of the many are considered' (Healey, 2012b: 62). Thus,

> the project is to change governance processes and cultures towards forms and practices within which critical, dialogic and discursive forms of deliberative democracy can flourish. The argument here is that such processes are more likely to promote attention to the values of social justice, environmental responsibility and cultural sensitivity than overly competitive processes and overly generalised ideologies.
>
> *(Healey, 2006: 317)*

With stress on 'how to make different voices heard in the collective decision-making process' (Fischler, 2000: 364), collaborative planning seeks to realise 'the intricate task of retaining allegiance to the utopian normative principles of the Habermasian project while also striving to gain credibility as a model of planning, a practice that is both capable of being carried out and socially worthwhile' (Harris, 2002: 41). In this way, collaborative planning shares with work by Forester, Innes and others a constructivist ontology, focus on the micro-dynamics of practice and concern for enhancing deliberative democracy through appeal to Habermasian-informed theorising. Accordingly, the critical pragmatism of Forester, Innes' attention to consensus-building and Healey's focus on collaborative planning are generally referred to under the umbrella term of 'communicative planning'.

Emerging in the 1980s from opposition to notions of planning as a scientific activity, 'by the mid-1990s, communicative planning was no longer just a concept. It was an intellectual perspective in planning theory, with a body of adherents, rubrics, philosophical referents, and a research agenda' (Healey, 2012a: 345). Gaining prominence during a period of post-positivist questioning (Allmendinger, 2002) and a growing call for marginalised groups to be given a greater voice in decision-making (Soja, 1996; Young, 1993), communicative planning offered promise to address many of the deficiencies of representation in planning activities. Indeed, it was during the 1990s that Sandercock (1998: 4) famously accused conventional forms of planning as being 'anti-democratic, race and gender-blind' and thereby supporting 'culturally homogenizing practices'. For her, use of the 'public interest' to legitimate planning decisions inevitably reflects the aspirations of the powerful. She holds that this can suppress the diversity of 'multiple publics' (1998: 197) found in the array of cultures and communities of contemporary cities. Sandercock thereby contends that planning is an 'always unfinished social project whose task is managing our coexistence in the shared spaces of cities and neighbourhoods in such a way as to enrich human life and to work for social, cultural, and environmental justice' (2004: 134). Yet, what Sandercock fails to realise is that while rejecting an explicit public interest (common good), she nevertheless tacitly endorses it in the 'social project' of the 'imagined Utopia' envisaged through her idea of 'Cosmopolis'. For example, when premising her argument, she describes Cosmopolis as 'a construction site of the mind, a city/region in which there is a genuine connection with, and respect and space for, the cultural Other, and the possibility of working together on matters of common destiny, a recognition of intertwined fates.' For Sandercock, 'the principles of this postmodern Utopia – new concepts of social justice, citizenship, community, and shared interest' require 'a new style of planning which can help create the space of/for cosmopolis' (Sandercock, 1998: 125). It is difficult to see how the principles of 'social justice,' 'citizenship,' 'community' and 'shared interest' in 'working together on matters of common destiny' speaks of anything but a sense of a shared good; a common good (or public interest). Indeed, Allmendinger (2017: 179) identifies how Sandercock is resonant with communicative planning theory through an assumption that allowing '"hidden" voices to speak will change existing processes and outcomes to the appeal to some overall concept of justice'. As such, Sandercock's view illustrates Flathman's (1966: 13) assertion regarding the idea of the public interest – namely, 'we are free to abandon the concept but if we do so we simply have to wrestle with the problems under some other heading.' Against this backdrop, the normative impetus of both cosmopolis and communicative planning appears to echo earlier endeavours of those wishing to envisage planning as 'advocacy' (Davidoff, 1965). However, whereas advocacy planning conceived planners as 'translating' community viewpoints into the technical jargon of bureaucracies, the planners envisaged both by Sandercock and communicative planning theorists 'let communities in need of planning assistance speak for themselves,' with communicative planning in particular directing attention to critically questioning 'the attitudes and procedures making it difficult for the disadvantaged to get their message through' (Sager, 2018: 95).

Nevertheless, it is on this very issue of giving voice to stakeholders that the concept of the 'common good' becomes foggy in communicative planning theory. Healey in particular is somewhat vague in this context. Her collaborative planning approach is generally ambiguous regarding the existence of a common good (or 'public interest'), stating that 'the culturally homogenous community with a common 'public interest' has been replaced in our imaginations by recognition of a diversity of ways of living everyday life and of valuing local environmental qualities' (Healey, 2006: 32). However, she still refers to a Habermasian-informed concept of the public interest that 'has to reflect the diversity of our interests and be established discursively' (p. 297). Sager is less elusive in contending that in communicative planning theory 'it might not be unreasonable to *define* the consensual outcome of dialogue as being in the public interest' (Sager, 2018: 97 – emphasis in original). This echoes the assertion by Innes and Booher (2015: 205): 'Although participants do not come into the process looking for the public interest, as they accommodate diverse interests, the proposals come closer to something that can be viewed as in the common good.' Such a line of reasoning has led some to conclude that 'whereas modernism held strong beliefs about project management and concrete outcomes defining good planning,' communicative planning theory presents 'no moral norm to shape the outcome, replacing it with a moral norm to shape the process' (van Dijk, 2021: 8). Thus, to conflate the outcome with the common good in communicative planning theory is to confuse means and ends in localised communicative planning endeavours with the broader democratic aspirations of the communicative planning project. Specifically, the impetus for communicative planning is a desire to address deficits in planning practice by realising the common good of an 'inclusionary ethic' that

> emphasises a moral duty to ask, as arenas are being set up, who are members of the stakeholder community? how are they to get access to the arena in such a way that their 'points of view' can be appreciated as well as their voices heard? and how can they have a stake in the process throughout.
>
> (Healey, 2006: 271)

Hence, the common good of communicative planning theory is the democratisation of planning.

Yet, this leaves somewhat problematic the position of the planner. As noted earlier, under communicative planning theory, the planner is reconceived as a 'knowledge mediator and broker' (Healey, 2006: 309). In this role, they are envisaged as seeking to advocate an 'inclusionary ethic' in advancing the common good. This is achieved by translating Habermas' 'ideal speech' concept into practice by way of 'collaborative rationality'. As profiled by Innes and Booher (2015: 209),

> The basic idea is that a decision can be collaboratively rational if it incorporates diverse and interdependent stakeholders who engage in authentic dialogue around a shared task. The ground rules of this dialogue establish expectations that participants will speak sincerely, be legitimate representatives of an interest, and provide accurate and comprehensible information.

Here, the planner engaged in deliberation is conceived as a person both seeking to, and capable of suspending prejudices in their role as a knowledge mediator and broker in an 'authentic dialogue' concerning a shared task, such as in making a decision on a divisive planning issue. This model thereby asks the planner to abstain from judgement and let others decide. How this is possible in the daily activity of planning is somewhat problematic, as it effectively requires that planners no longer engage in the act of seeking to instantiate their evaluation of circumstances. As noted by Allmendinger (2017: 264), it

> questions the whole basis of a 'planning profession' – how can you have a profession (whose *raison d'être* is the application of expert knowledge) if you argue that there is no such thing as expert knowledge, only different opinions to be brought together?

Furthermore, others argue that reconceiving the planner as a knowledge mediator and broker runs against real-world practice constraints such that there arises 'the possibility that planners are not able to act as disinterested mediators, exactly by virtue of their roles in relation to the state' (Huxley, 2000: 375). Hence, while communicative planning theory is contoured by a Habermasian-informed desire to address distortions of power in planning practice (Forester, 1989, 1993; Healey, 2006; Innes and Booher, 2010), critics contend that the problematic of power asymmetries are an indelible feature of governance. Forester (1993) appears to acknowledge this[5] but is stymied in formulating an effective response by communicative planning theory's primary focus on the micro-dynamics of interaction at the expense of the structural forces that may shape those interactions.

Indeed, this criticism is one of the main issues on which Susan Fainstein formulates her perspective in contradistinction to communicative planning theory (Fainstein, 2010). Fainstein's work has evolved over several decades from Marxist critical political economy to a focus on 'justice'. Her contemporary work is differentiated from communicative planning by 'its emphasis on material distribution and substantive outcomes, over deliberation and inclusive participation' (Campbell et al., 2014: 48). Equally critical of 'postmodern calls for diversity as an unquestioned orthodoxy' (Connolly and Steil, 2009: 7) and the simplification of difference by Marxism through an overriding concern with socio-economic class,[6] she proposes a reading of justice that incorporates the values of democracy, diversity, and equity. Fainstein believes that each of these exists 'in tension with each other' such that 'maximizing any of them can result in trade-offs' (Fainstein, 2018: 130). She holds that the communicative planning approach is faulted by two fundamental constraints: firstly, its proceduralist focus fails to take account of structural inequalities and hierarchies of power that hinder the search for consensus (Fainstein, 2000); secondly, the primary focus on process inhibits effective evaluation of substantive outcomes and thereby cannot guarantee the realisation of justice in the decisions made (Fainstein, 2005). Accentuating the substantive dimensions of planning, and with an eye on the issue of scale, Fainstein proposes an alternative 'Just City' model which 'subsumes the communicative approach in that it is concerned with both processes and outcomes but also recognizes the potential for contradiction between participation and just outcomes.' Ultimately, she holds that 'just outcomes should trump communicative norms when the two conflict' (Fainstein, 2016: 261). Thus, 'for just-city theorists the principal test is whether the outcome of the process (not just of deliberation but of actual implementation) is equitable; values of democratic inclusion also matter, but not as much' (Fainstein, 2010: 10). Fainstein thereby agrees with the need to enhance the voice of affected publics. However, she contends that the more open, more democratic process that concerns communicative planning

> fails to confront adequately the initial discrepancy of power, offers few clues to overcoming co-optation or resistance to reform, does not sufficiently address some of the major weakness of democratic theory, and diverts discussion from the substance of policy.
>
> *(p. 24)*

As such, her

> criticism of the proceduralist emphasis in planning theory is not directed at its extension of democracy beyond electoral participation but rather at a faith

> in the efficacy of open communication that ignores the reality of structural inequality and hierarchies of power.
>
> *(p. 30)*

Fainstein considers that 'there is a naïveté in the communicative approach, in its avoidance of the underlying causes of systemic distortion and its faith that reason can prevail' (p. 34). In response, she develops an 'evaluative standard by which to judge urban policies and to express the goals of urban movements' (p. 36). This involves using case study material as exemplars for extracting a set of norms and criteria that can be wielded in maximising equity, diversity and democracy. Hence, Fainstein criticises communicative planning by identifying the limitations of its approach and subsuming its objectives within a more expansive concept of good planning conceived through the lens of social justice (Fainstein, 2009). In this sense, her conception of the planner differs from the mediator-broker of communicative planning. Echoing Krumholz et al. (1975: 170), she argues that 'planners whose aim is justice need to intervene in the planning process, calling for policies that favour low-income and minority groups' (Fainstein, 2014: 8). Thus, Fainstein's sensitivity to the broader dimensions of power leads her to support the democratic ideals of communicative planning but criticise what she views as its over-emphasis on distortions within the 'processes' of deliberation rather than the broader social inequities that fuel such distortions.

In essence, Fainstein proffers a substantive theory of justice in planning as a good that is grounded in an appeal to advance equity. This is reasoned from within a perspective on planning drawn from exemplars. While presenting a coherently argued understanding of planning and advocating a concept of what justice entails with respect to this, the model of the planner that she posits is somewhat lacking. This is because the planning subject of the Just City perspective is one whose reflective capacity is curtailed by how the argument is constructed. Here, the key vectors positioning what is right/wrong and better/worse are specified prior to the planner entering the decision arena. Furthermore, the respective weights given these are stipulated such that equity trumps democracy and diversity when trade-offs are required. As such, this perspective subjectifies the planner in a way that reduces reflective agency by predetermining what 'really' matters and why. Yet the very differences of opinion on what matters that are advanced by different theoretical perspectives in the review contained in this and the next chapter, suggests that there are different ways of reasoning what counts. Ultimately, the manner in which the Just City argument is profiled is thereby problematic for an appreciation of how the commonness of the common good supplies a legitimation for planning. Therefore, while the Just City perspective advances a well-considered understanding of planning and the substantive concept of justice in planning, its neglect of the planning subject as a complex agent with moral concerns that may depart from the perspective's preordinates, undermines the potency of the perspective for clarifying how planning can have a common good in a world of different viewpoints on what this good may be.

A Conceptual Conundrum

Although communicative planning theory emerged in response to the perceived hubristic pretensions of framing planning as a scientific activity, it nevertheless shares with both systems theory and the rational process approach a view that planning is a force for good, that the goods of planning can be identified and that these goods are typically common in nature. These otherwise divergent approaches also share a deep concern with proper process, albeit what constitutes 'proper' and 'process' in each case may differ significantly. Similarly, while fundamentally disagreeing on the model and role of the planner, these perspectives nonetheless share an appeal to the even-handedness of the professional planner as a necessity for progress. Specifically, in systems theory the planner is constituted as 'the expert' who is best placed by way of access to data and training to determine what the common good entails. In contrast, the planner of the rational process approach eschews normative debate on what 'ought' to be done in favour of 'how' to deploy their technical expertise to inform and help realise the common good identified by others. Drawing on pragmatic and Habermasian philosophy, as well as insights gleaned from anthropology and institutional theory, communicative planning theorists reject the exclusionary tendencies of such appeals to scientific expertise. Instead, they highlight the need for greater inclusivity as both the normative and practical basis for advancing planning as a force 'for good' by identifying what stakeholders determine as 'the good'. Here, the planner is but one agent navigating conditions criss-crossed with contention, compromise and consensus 'in the flow of life lived in association with others' (Healey, 2009: 287). Each person brings their own range of expertise, opinions and experiences in helping to identify a mutually agreed path forward, which is normatively considered the best, even if not the most efficient and effective in meeting the challenges at hand.

It may therefore be contended that communicative planning theorists have developed a subtle understanding of the planner. However, these theorists weave complex interpretive perspectives into the constitution of the planner only to produce an inconsistency in the role of the planning subject. In this they advance readings of the planner as a context-situated agent seeking to negotiate their 'institutional' embeddedness in time and space. But they then fashion the planner as one who seeks to achieve a non-distorted form of planning practice yet must at times actively distort it by using their professional position to advocate on behalf of and/or reshape the deliberative arena to ensure that voices 'they' perceive as marginalised are given a hearing. In this vein, Tewdwr-Jones argues that as Forester (1999b) 'believes planners should act in the interests of the disadvantaged or oppressed,' communicative planning theory cannot achieve a non-distorted form of planning practice because

> the planner under Forester's thesis is expected to enter into the arena with a pre-packaged notion that he or she is there to act for a particular interest or need; in other words, the planner enters with an *a priori* assumption that their role is to facilitate the disadvantaged.
>
> *(Tewdwr-Jones, 2002: 72 – emphasis in original)*

Hence, the picture of the planner painted is of one divided between receding into the background once 'they' have determined that correct deliberative conditions are met and intervening once 'they' decide that deliberative conditions need amendment. Consequently, the planner is allocated considerable power to frame and shape deliberations through their role in 'organising attention' (Forester, 1989) – agendas, participants, ground rules, etc. Even in cases where such framing and shaping are ostensibly performed in a collaborative format, it is the planner who has the power to initiate the process and influence where and when it will take place, as well as affect the determination of who should participate in the deliberative exercise. Still, the planner is nonetheless most frequently portrayed as simply the midwife to others' decisions; bound by a 'discursive ethics' in 'protecting and nurturing voice' (Forester, 1994: 202) through acting as facilitator, moderator and mediator wholly concerned with dialogue, debate and negotiation (Forester, 2009). This leaves the planner as a split subject splintered between neutrality, partiality and the ineradicable influence of their own views on what should be done, by who, where, when and why, while all the time wrestling with their normatively pre-defined role as impartial 'knowledge broker and mediator' (Healey, 2006). Ultimately, the problem here may lie not so much in the normative objectives of communicative planning as in the conceptual conundrum created by how it has attempted to work these out.

In essence, striving to reconcile the paradox of 'partial impartiality' at the heart of the approach generates confusion between substantive and procedural perspectives on how to advance a shared sense of the good. This dilemma both constitutes and is consequent on the relationship between the planning subject and the concept of the common good captured by the approach's 'inclusionary ethic' (Healey, 2006: 271). As realised by Fainstein in her Just City perspective, resolution is only possible if either the substantive or procedural perspective takes priority. Hence, implicit in communicative planning theory is the primacy of a substantive commitment to removing asymmetries of power to enhance democratic inclusion in decision-making. This effort underpins the theory's view on the raison d'être of planning as a force for good (Healey, 1992b). Accordingly, it is this substantive concept of the common good of inclusive democracy that supports the focus on the procedures through which the explicitly consensual common good is identified (Innes and Booher, 2015; Sager, 2018). Thus, as with systems theory, the rational process approach and the Just City perspective, the effort to formulate and advocate communicative planning theory thereby represents a moral position located at a juncture in the discipline's evolution that is born of intersubjective reflection by theorists and practitioners on what good planning should serve and what being a good planner involves. Grasping this requires an appreciation of the co-constitutive relationships between planning, the planner and the concept of a common good. To omit or prioritise any one of these is to gain only a partial understanding of how each is constituted. Part 2 of this book is thereby concerned with (re)conceiving these relationships. But before we proceed with this, reviewing the illuminations and elisions supplied by a contradistinctive family of 'critical' perspectives is first necessary to appreciate the variety of theoretical understandings, insights and deficiencies as to what planning comprises, the feasibility of a common good and how the planning subject is comprehended.

Notes

1 Friedmann was to develop his ideas into more politically radical forms of transformative planning in later years (Friedmann 1987, 2011).
2 In an early critique of Forester's approach, Campbell and Marshall (1999: 474) contend that 'despite its radicalism . . . his approach is still very much set within the Rawlsian tradition. It is concerned not so much with *substantive ends* but with how planners can intervene to counteract imbalances of power in an effort to create a "fairer" process within which conflicts can be mediated and resolved' (emphasis in original).
3 Sager (2012) has since sought to develop an approach that is more critically attentive to the distorting role of neoliberalism in planning, wherein he theorises a number of responsibilities that academics must hold with respect to theoretical development and teaching, as well as towards advancing an inclusionary ethic.
4 Mattila (2016: 355) identifies that 'Healey speaks of "localized lifeworlds", emphasizing the diversity of lifeworlds, whereas Habermas is mainly interested in the universally shared structures of our lifeworlds, structures that he derives from the rational reconstruction of the presuppositions of argumentative speech' (Healey, 2006, 1987; Healey, 1997: 62).
5 Moreover, there is a sense that Forester became somewhat frustrated with academic commentary on the existence of power distortions in planning rather than action in seeking solutions to address an identified problem. This is intimated when he asserts, 'Lets us stop rediscovering that power corrupts, and let's start figuring out what to do about the corruption' (Forester, 1999b: 9).
6 In this context, Fainstein (2014: 12) has maintained a long running but respectful debate with Marxists thinkers, noting, for example, 'An important difference between my view and David Harvey's is that I am willing to embrace reform through existing political-economic processes rather than viewing greater justice as unattainable under capitalism.'

2
A CRITICALLY UNCOMMON GOOD

Critical Planning Theory

Although an amorphous array of approaches that contend as much as cohere, an evolving range of planning theories can be loosely associated with a 'critical' focus on questions of power that are held as the central consideration in understanding planning and effecting positive change. Theorists of this genus examine a wide variety of issues to reveal the processes through which privilege and marginalisation are shaped and distributed in policy and practice. Such theorists often seek to provide a reflective check on planning theory and practice by unveiling ideological assumptions and limning the means through which distortions emerge and embed in modes of governance. Frequently this probing and exposing is accompanied by specifying tactics of resistance and trajectories for change. Therefore, an initial pass may suggest that such critical approaches share with communicative planning theory the premise of a shared sense of the common good. However, an important supposition of critical perspectives is that the interweaving of politics and power to generate advantaged positions means that the idea of a 'common' good is fundamentally problematic. To an extent, the political economy of Marxist-influenced theorists (Dear and Scott, 1981; Fainstein and Fainstein, 1979; Harvey, 1978) provides a percussor to some of this work and continues to generate interest in a less structurally determinist fashion (Marcuse et al., 2009). However, more potent influences on modern thinking have come from the work of continental philosophers and social theorists ranging from past luminaries such as Nietzsche and Foucault to towering contemporary figures like Mouffe and Rancière. Each provides inspiration for a different sketch of planning's deficiencies and helps critical theorists chart a normative concept of what 'better' planning may mean, albeit this can at times be implicitly delivered. While each of these lineages follows its own course, they all postulate that a critical approach to planning can

DOI: 10.4324/9781003155515-4

reveal asymmetries of power, which if addressed can advance a better form of planning. This presupposes that the revelations of critical perspectives have a shared (if not universal) benefit. Hence, while critical perspectives in planning theory have an uncomfortable relationship with the idea of a common good, this relationship is immanent to the perspectives themselves and bound up with the normativity of the positions taken on what is right and wrong. What emerges as implicit to such perspectives is thereby the degree of commonness of this good, rather than the view that it is somehow explicitly shared. Early forays into this terrain emerged as theorists sought to uncover the stealthy manifestation of power operating through conceptions of planning as a 'rational process' and the pretensions of communicative planning theory as a means to redress issues of distorted representation. To the fore of such work is a Foucauldian-inspired defenestration of planning as an impartial process of decision-making.

The Foucauldian Critique

Work by French philosopher Michel Foucault displays a social-constructivist ontology that is historicist in orientation, although his work is frequently applied to contemporary issues. Foucault set himself the project of destabilising presumptions of 'the given' in social relations. In this context, 'uncritical acceptance of anything presented as natural, necessary, or ineluctable is problematic from a Foucauldian perspective' (Taylor, 2011: 4). He holds that such uncritical acceptance permits the emergence of power relations that assist the perceived legitimation and dominance of a restricted array of reasoning and behaviour. As a consequence, alternative modes of thinking and doing are rendered invalid or immoral, thereby warranting social sanction and suppression. Thus, what Foucault shows is that the current 'order of things' (Foucault, 2002) is not inevitable. Instead, it is contingent and so could be otherwise. His project is thereby to unmask the 'ontology of the present' (Foucault, 1986: 96). Accordingly, he explores 'the limits of ways of thinking to find possibilities for thinking differently' (Cooper and Blair, 2002: 513). Foucault endeavours to reveal the contours of these limits and expose their effects by drawing attention to how 'the subject' is constituted. By proposing that subjects 'are made' (Foucault, 1982: 777), Foucault challenges the long-held view in western philosophy extending from the Enlightenment through to the present that the subject inherently possesses an indelible 'substance' which endows them with agency by means of an objective rational faculty (Foucault, 1987). He thereby seeks to demonstrate that the subject is shaped by and transmits socio-cultural norms. In this sense, a Foucauldian sensitivity to the subject acknowledges the role of context in forging our understanding of the world. Central to this perspective is attention to the part played by 'power'.

A Foucauldian approach holds that conduct is governed by the emergence of relations of power carried via forms of rationality that mould perceptions in ways that privilege and marginalise different people and ideas by producing subjectivities (Lennon and Fox-Rogers, 2017). Thus, to appreciate the effects which flow from the way subjectivities 'are made,' one must remain aware of the role played by power in the formation of such subjectivities. From a Foucauldian perspective, 'Power should be seen as a verb rather than a noun, something that does something, rather than something which is or which can be held onto' (Mills, 2003: 35). His approach suggests that power is 'immanent in' (Foucault, 1990: 94) all social relations. As such, a Foucauldian view holds that power 'is "always already there," that one is never "outside" it' (Foucault, 1977: 141). Much of Foucault's work is concerned with tracing the historical constitution of this power and the way such constitution creates asymmetrical relations that produce subjectivities. His work on power, at least in its earlier manifestations, thereby involves borrowing from Neitzsche the 'geneological' method of unearthing the provinance and passage of concepts through time and space. Hence, while to Forester (1989, 1993) power asymmetries represent 'distortions,' for Foucault power asymmetries are simply an indeliable attribute of human interaction. Indeed, Foucault (1990: 7–8) contests the 'repressive hypothesis' that regards power solely as an oppressive force curbing liberty. In its place, he advances a nuanced understanding of power that sees it as concurrently 'productive, something which brings about forms of behaviour and events rather than simply curtailing freedom and constraining individuals' (Mills, 2003: 36).

Nonetheless, early uses of Foucault's work in planning academia emphasised the repressive dimensions of power. For example, Christine Boyer (1983) laced a Foucauldian genealogy through a Marxist diagnosis to produce a critical history of American city planning in the first half of the twentieth century. Boyer emphasised Foucault's work on prisons (Foucault, 1991) to reinterpret the function and operation of planning as a 'disciplinary technology' deployed to resolve the discord between the need for order and the laissez-faire proclivities of capitalism. For Boyer, the attempts by planners to gain control of urban trajectories through the ordering of space were only partially successful, as too much regulatory interference was seen to hinder opportunities for profit generated by escalating land prices, with urban congestion then resulting from the permissive regulatory environment created. Such a Foucauldian reading of urban planning helped foreshadow work by academics such as Roweis (1983), Huxley (1994) and Harris (2011), who have explored the 'dark side of planning' as manifested in the institutionalisation of oppressive planning practices (Yiftachel et al., 2001). Yiftachel in particular has focused on the nefarious aspects of planning that he contends subvert its ostensibly progressive aims (Yiftachel, 1998: 275). However, mobilising a 'Lacanian insight and a dash of Derridean deconstruction,' Allmendinger and Gunder (2005: 97) criticise the binary opposition posited by Yiftachel in his earlier work as a 'simplistic bifurcation of good and "bad"' which serves to create 'a neat dualism that presents a black and white image of "acceptable" and "unacceptable" planning practice but clearly ignores the messiness and undeciderabilitiy (sic) of reality.' In their more mature work, Yiftachel (2009) and others researching the 'dark side of planning' (Fox-Rogers and Murphy, 2014; Watson, 2003) adopt a less dichotomising perspective on planning, which is more attentive to the nuances of practice but nonetheless seeks to expose the shadier dimensions of planning's underbelly. Much of this work has examined the deployment of state-sanctioned rationalities to facilitate questionable practices that marginalise certain actors in facilitating political objectives (Fox-Rogers, 2019; Lennon and Waldron, 2019).

This line of research is exemplified by Flyvbjerg (1998) in his influential study of transport controversies in Aalborg, Denmark. Operating within the ambit of 'dark side' research (Flyvbjerg, 1996, 2002), Flyvbjerg's Aalborg study traced a lineage of thinking from Machiavelli through Nietzsche and onto Foucault in analysing how ideals may be warped in application through the power-infused operation of ('real') planning rationality (*Realrationalität*). Here, planning activity is understood as open to corruption as backstage politics delineate the forms of rationality ostensibly employed by planning. This leads Flyvbjerg (1998: 234) to conclude that 'while power produces rationality and rationality produces power, their relationship is asymmetrical. Power has a clear tendency to dominate rationality in the dynamic and overlapping relationship between the two.' It is therefore unsurprising that Flyvbjerg's research provokes a rather dubious faith in the concept of a 'common good' or 'public interest'. As he notes with respect to Aalborg, 'Institutions that were supposed to represent what they themselves call the "public interest" were revealed to be deeply embedded in the hidden exercise of power and the protection of special interests. This is the story of modernity and democracy in practice, a story repeated all too often for comfort for a democrat' (Flyvbjerg, 1998: 225). Such effort at unearthing the 'hidden exercise of power' subsequently leads Flyvbjerg and Richardson (2002: 44) to question the usefulness of communicative planning theory by drawing on the 'analytics of Michel Foucault' to ask, 'what is actually done,' as opposed to Habermas's focus on 'what should be done'. Flyvbjerg and Richardson hold that 'Foucauldian analysis, unlike Habermasian normativism' offers planning theory 'better prospects for those interested in bringing about democratic social change through planning' (p. 45). In their view, it is thereby possible to negotiate a path out of planning's dark side by unveiling and questioning the operation of power in planning, rather than ignoring its inherent presence. This distinct power-aware approach to planning has led Flyvbjerg (2004) to suggest that the planner refocus on himself or herself as the object of concern. The normative emphasis of this 'phronetic' approach is on determining how power asymmetries exert a restraining effect on people, with an associated recommendation that planners reflect on what they ought to do in light of such knowledge. While this work has done much to illuminate the role of power in planning practices, it conceives planners as operating in situations where they are invariably aware of the power dynamics at play (Lennon and Fox-Rogers, 2017). As such, it paints a picture of planning as a place where practitioners can be held morally culpable for marginalising certain interests in the planning arena. Certomà (2015) has helped address this issue by demonstrating the value of expanding Flyvbjerg's approach via the Foucauldian concept of 'governmentality'. Broadly conceived, this concept addresses 'How we think about governing others and ourselves in a wide variety of contexts' (Dean, 2010: 267). It refers to the different rationalities or 'mentalities of government'

(Rose and Miller, 1992) that guide perspectives on thinking and doing. By focusing on how diffuse forms of control inform popular mentalities (Pløger, 2008) and by expanding the concept of government beyond the classically conceived 'state,' it advances a more subtle understanding of how a variety of influences emanating from an array of sources shape individual and communal behaviour (Certomà, 2015; Huxley, 2006). Others have employed Foucault's concept of governmentality to critique neoliberalism as an entrepreneurial reconfiguration of planning in the context of market rationalities to produce consumerist neoliberal spaces and subjects (Davoudi and Madanipour, 2013; Raco and Imrie, 2000; Vanolo, 2014). As summarised by Huxley (2018: 209),

> Very broadly speaking then, in those studies seeking to draw attention to negative implications of planning, the main uses made of critical Foucauldian concepts have been to highlight aspects of the exercise of state and/or professional technical power, whether as political *realrationalität*, as disciplinary regulation and surveillance or as governmental redirection of the rationales for planning and the neo-liberal subjectification of those involved.

Indeed, the conventional deployment of Foucault in planning research has largely focused on exposing the adverse dimensions of power on practice. However, this bank of work has tended towards the evacuation of agency such that the planner is compressed into an actor whose subjectivity is determined by the pervasive forces of power that circulate in, and structure society through historically specific totalities of discourse and practices, totalities which Foucault (1977) refers to as 'dispositifs'.

Yet, an emerging line of research has sought to explore the more 'constructive' dimensions of Foucault's oeuvre and thereby help reinvest the planning subject with agency. Huxley (2018) for example argues for an exploration of Foucault's concept of 'counter-conducts' that provide openings for transformations in planning by provoking and responding to governmental subjectifications as an active form of critique. In a similar vein, Inch (2012) appropriates and broadens the concept of 'cultural work' from the sociology of the professions to examine the variety of practices which actors use to negotiate a given power-infused context. He views this as 'broadly compatible with a Foucauldian concern for the micro-practices that emerge at the sites where forms of normalising or disciplinary power meet the lives of those they target and are either incorporated or reworked' (Inch, 2018: 201). For Inch, attention to cultural work as modes of counter-conducts 'highlights the presence of numerous sites where actors, individually or collectively, more or less actively, negotiate a sense of self and fit with prevailing regimes, sometimes in ways that reshape the regime itself.' Hence, he believes that a part of the challenge of understanding the scope for better planning is to 'examine under what circumstances cultural work might coincide with what Foucault called the "work of freedom" and the exercise of agency to assert alternative possibilities' (p. 201). Lennon and Fox-Rogers (2017) have specifically focused on this dimension of Foucault's work, noting Foucault's direction that his later work on ethics be used to interpret his earlier analysis of power and knowledge (Flynn, 2010). For Foucault, the concept of 'ethics' differs from that which is conventionally conceived in moral philosophy as either the study of abstract ethical models or the scrutiny of normative criteria for applied action. Instead, Foucault defines ethics as 'the kind of relationship you ought to have with yourself' (Foucault, 1984a: 345). He perceives ethics as comprising 'those intentional and voluntary actions by which men not only set themselves rules of conduct, but also seek to transform themselves, to change themselves in their singular being' (Foucault, 1984b: 10). In this sense, Foucault's conception of ethics comprises a self-forming activity of moral constitution that must occur in a contingent social context inherently infused with power/knowledge. Lennon and Fox-Rogers (2017) deploy this understanding to identify ethical issues arising from the approaches used by practitioners to justify their planning activities. While such work has helped reinfuse a Foucauldian sensibility with agency, the connection between such agency and a conception of the common good remains tenuous at best. Indeed, as successors to the 'school of suspicion'[1] (Ricoeur, 1970), the idea of a common good is problematic for Foucauldian scholars. This is because 'counter-conduct,' 'cultural work,' 'ethics of the self' and other subject-centred tactics of resistance informed by Foucault's (Nietzschean-influenced) thought are focused on negotiating the constraints set on one by the subjectifying forces circulating in society. The 'common' is thereby challenged as it would invariably constitute another dimension of those forces that the tactics of resistance are seeking to contest. As noted by Huxley (2018: 217),

> Engaging with the implications of counter-conducts is a difficult position for those who defend planning as a besieged guardian of the 'public good' to accept . . . planning seeks to conduct the conduct of others and thereby incites counter-conducts of refusal.

Hence, emphasising difference becomes an important aspect in resisting the potential annihilation of agency that acquiescing to a collective 'common good' may entail. Beyond the historicist constructivism of Foucault, such misgivings on the latent tyranny of the 'common good' concept is a feature shared across much political philosophy and social theory (Arendt, 1951; Laclau and Mouffe, 1985; Žižek, 1989), some of which has influenced the growing stream of 'post-political' thinking in planning theory.

The Post-Political Critique

Unlike movements such as communicative planning theory, the post-political critique does not amount to a clearly identifiable theory of planning. Rather, it distinguishes a shared critical sensitivity. As noted by Allmendinger (2017: 196),

> Post-political thinking emerged in part to explain the contradictions between, on the one hand, the outward shifts in planning towards more open,

transparent and supposedly consensus-building approaches and, on the other hand, the simultaneous rise in opposition to development, conflict, antagonism and distrust of government and officials, including planners themselves.

The post-political 'critique' thereby encompasses a range of work in planning academia that seek to expose and question the ways in which governance innovations ostensibly designed to promote participation and consensus can produce democratic deficits rather than address them (Allmendinger and Haughton, 2015; Legacy, 2016; O'Callaghan et al., 2014; Raco, 2014). For this reason, post-political studies endeavour to determine and unveil new forms of politics in planning processes that serve 'to supress or pre-emptively foreclose possibilities towards dissent through by-passing, short-circuiting or circumscribing difficult and contentious issues concerning the specification of desirable futures' (Metzger, 2018: 181). Such work is generally constructed on a philosophical platform developed throughout the late 1980s and 1990s by post-foundationalist political thinkers such as Chantal Mouffe and Jacques Rancière.[2] For Mouffe in particular, the malaise of contemporary forms of governance in modern western democracies is ontologically grounded (Merchart, 2007). Working with Ernesto Laclau, Mouffe brought together the thinking of theorists such as Lacan, Derrida and Gramsci, among others, to forge a political philosophy that is fundamentally defined by an ontology of 'lack' (Hillier, 2003). For Laclau and Mouffe 'every identity is dislocated in so far as it depends on an outside which both denies that identity and provides the condition of possibility at the same time' (Laclau, 1990: 39). Consequent to this 'constitutive outside,' there are no fixed identities. Rather, identities are formed through 'articulation,' which is the practice of 'establishing relations among elements such that their identity is modified as a result of the articulatory practice' (Laclau and Mouffe, 1985: 105). Articulation involves a decision in favour of a particular sign-signified association over other possible associations that cannot be justified with reference to some transcendental order of objectivity beyond the process of the association itself. In this sense, what is perceived as common sense or proper is just a contingent set of social relations, which in keeping with Foucault is the space where power is to be found. Hence, Laclau declares, 'objectivity – the being of objects – is nothing but the sedimented form of power, in other words a power whose traces have been erased' (Laclau, 1990: 60). Accordingly, Laclau and Mouffe follow Derrida (1978) in rejecting the essentialism of approaches that seek to objectify ideas such as 'class'. Working with this anti-foundationalist framing, they instead extend the class-orientated concept of hegemony proposed by Gramsci (1971) to formulate an understanding of how articulatory practices function to normalise particular constellations of meaning in ways that privilege and marginalise different actors in a field of potentially contesting interpretations. As Mouffe (1999: 756) argues, 'Every consensus exists as a temporary result of a provisional hegemony, as a stabilization of power and that always entails some form of exclusion.' Thus, by exposing the contingency and temporality of dominant ways of relating, they reanimate the agency of actors by highlighting that any apparently fixed articulation can be re-articulated and thereby changed. At first blush, Mouffe's work may therefore appear to promise a nuanced conception of the subject that can help inform a refined understanding of the planner. However, as argued next, Mouffe's views become problematic for post-political thinkers in planning theory consequent on how she develops her perspective.

Core to this perspective is the idea of 'antagonism' that 'constitutes the limits of every objectivity, which is revealed as partial and precarious objectification' (Laclau and Mouffe, 1985: 125). Mouffe expands upon this seam of thought in her later works. Here she constructs a radical democratic theory of 'agonistic pluralism' in contradistinction to the deliberative democratic view that underpins communicative planning theory. In keeping with her earlier Lacanian-informed ontology of 'lack'[3] and Derridean-infused concept of the 'constitutive outside,' Mouffe's interpretation is devised through a reworking of inclusion-exclusion dualisms from Carl Schmitt's thinking on what comprises 'the political' (Schmitt, 2008 [1932]). However, rather than extrapolating from this that there is no place for pluralism within a democratic political community – as Schmitt had done – Mouffe thinks 'with Schmitt against Schmitt' (Mouffe, 2005: 14) to emphasise the importance of difference in reconceiving what constitutes 'the political' and 'politics'. As she explains,

> By 'the political' I mean the dimension of antagonism which I take to be constitutive of human societies, while by 'politics' I mean the set of practices and institutions through which an order is created, organizing human coexistence in the context of conflictuality provided by the political.
>
> *(Mouffe, 2005: 9)*

For Mouffe, 'the political' is the antagonism that is 'inherent to all human societies' (Mouffe, 2013: 2), which is characterised by an absence of respect for the differing opinions of others. She believes that this is the space of unproductive contestations between those framed as 'enemies'. In contrast, 'politics' seeks to domesticate 'the political' through 'the ensemble of practices, discourses and institutions that seek to establish a certain order and to organise human coexistence' (p. 2–3). Mouffe believes that it is 'the lack of understanding of 'the political' in its ontological dimension' that lies at the origin of 'our current incapacity to think in a political way' (Mouffe, 2005: 9). In her view, 'Too much emphasis on consensus, together with aversion towards confrontations, leads to apathy and to a disaffection with political participation' (Mouffe, 2013: 7). Consequently, she argues that 'Democratic politics cannot be limited to establishing compromises among interests or values or to deliberation about the common good; it needs to have a real purchase on people's desires and fantasies.' To Mouffe, this means that 'democratic politics must have a partisan character' (Mouffe, 2005: 6). The challenge is thereby how to render such passionate partisanship constructive. To achieve this, she draws on the Nietzschean (2009 (1872)) and Arendtian (1958) notion of 'agon'. This term stems from ancient Greece and refers to the inevitable struggle of social and political engagement that can be cultivated as a civic virtue. Here, antagonism is tamed via mutual respect and reciprocity into 'agonism' (Mouffe, 2000). Mouffe explains this by describing how

> while antagonism is a we/they relation in which the two sides are enemies who do not share any common ground, agonism is a we/they relation where the conflicting parties, although acknowledging that there is no rational solution to their conflict, nevertheless recognize the legitimacy of their opponents.

Agonism thereby reframes the 'enemies' of antagonism into 'adversaries' such that their differences play out through 'politics' as they 'see themselves as belonging to the same political association, as sharing a common symbolic space within which the conflict takes place.' It is in this context that Mouffe believes that 'the task of democracy is to transform antagonism into agonism' (Mouffe, 2005: 20). Whereas divergent positions are thus facilitated within agonism, a meta-consensus (Dryzek and Niemeyer, 2006) on a common commitment to the principles of 'liberty' and 'equality' is central to a productive politics, even though disagreement is to be expected on how these principles are interpreted. Hence,

> What exists between adversaries is, so to speak, a conflictual consensus – they agree about the ethico-political principles which organize their political association but disagree about the interpretation of these principles.
> *(Mouffe, 2013: 318)*

Mouffe believes that 'far from jeopardizing democracy, agonistic confrontation is the very condition of its existence' (Mouffe, 2005: 30). In this way, a tacit sense of the common good is smuggled in through the backdoor. Considered in the context of planning, this implies ensuring that democratic decision-making processes advance liberty and equality as foundational in legitimating the democratic credentials of the decisions made. Thus, the substantive commitment to these definitional dimensions of democracy become the common good that profiles how democratic decision-making in planning should be arranged. This desire to maintain conflict within a public sphere that not only acknowledges it but sees it as supporting democracy is thus different to the consensus-seeking objectives of communicative planning theory, which is rooted in Habermasian political philosophy (McAuliffe and Rogers, 2019). This poses the challenge of 'how to develop agonism in planning into a proactive and pragmatic planning theory and not just a theory of agonistic democracy in the context of planning' (Mouat et al., 2013: 164). This conundrum stems in large part from the fact that like the thinking of Habermas and Rawls, Mouffe's thesis targets the constitutional level of politics rather than the particularities of planning practice. Hence, in attempting to translate her theory into planning, theorists remain 'somewhat sweeping and vague' (Metzger, 2018: 186) as to the discipline's role in dealing with conflict in agonistic conditions. As queried by Kühn (2020: 5), 'is it about resolving such conflict or, on the contrary, about escalating confrontation? Should conflict be "negotiated", "moderated", "mediated" or "arbitrated"?' Likewise, the role of the planner is somewhat unclear in agonistic planning. In this sense, Pløger (2004: 83) notes how 'The planner as a civil servant is placed within an agonistic field constituted by conflicts and disputes about what has been said, what is not said, who said what and when, and so on.' Yet agonistic planning theory lacks clarity as to how the planner should be conceived within this 'field'. For some, the post-political critique 'highlights the changing role of the planner from a progressive function, underpinned by agreed values and apolitical self-perception, to a narrower, partisan, pro-growth and development sensibility' (Allmendinger, 2017: 192). However, agonistic planning theory largely fails to furnish a considered alternative. Hence, 'the particular role attributed to planners remains rather unclear in the agonistic model. There is little concrete information on what "conflict management" . . . means in practice, or whether planning has the ability to deal with conflict productively' (Kühn, 2020: 9). In this context, role attributions such as 'wandering planner,' 'editorial organizer of dialogues' and 'non-excluding strategic navigation' (Pløger, 2018: 273) do not appear convincing. This leads Kühn (2020: 9) to deduce that the role for planning, and thus the planner, is simply 'to offer public arenas for disputes'.

Many of these issues also carry across to planning theory influenced by the philosophy of Jacque Rancière, even though Rancière's reading of post-politics is somewhat different to Mouffe. While Mouffe contends with broad theoretical issues on the agonistic constitution of modern democracy, Rancière examines how equality is denied or achieved in specific contexts. As noted by Legacy et al. (2019: 276),

> Rancière's point of departure from Mouffe is in the treatment of politics. He sees democracy as less a phenomenon that views politics through the prism of adversaries, but instead argues that politicisation is not attached to identity but is predicated upon claims for equality which can be episodic, but no less powerful.

The originality of Rancière's thinking is betrayed by his theoretical vocabulary. As with many twentieth-century French philosophers of his generation, Rancière begins with ideas that his audience could be expected to have some acquaintance with before swiftly de-familiarising the terms he employs as a means to confront assumptions and shed light on the processes he wishes to discuss. An example of this is his idiosyncratic concept of 'equality'. Rancière's notion of equality differs from most conventional interpretations of the term. This is resultant from his elision of standard debates divided by positions on liberty and redistribution. Indeed, Rancière recasts this debate in the context of what May (2008) refers to as 'passive' and 'active' stances on equality. May argues that while equality is differently conceived in the work of libertarian theorists such as Robert Nozick, liberal political philosophers such as John Rawls and Amartya Sen or even neo-Marxist thinkers like David Harvey, they are united by a 'passive' conception of equality that 'concerns what institutions are obliged to give people' (May, 2008: 4). Hence, from this 'passive' position, the equality sought for people 'is not something they create;

it is not something they guide; it is not something they do', such that while they may use the equality they enjoy in a multitude of ways, 'equality itself comes to them (or is protected for them) from a source outside of themselves' (May, 2010: 70). However, for Rancière, equality is 'active' in that it is presupposed as already existing among all participants in any situation (Davis, 2013). This is consequent on the equality of capacities by all participants in a hierarchy to perceive their position in that hierarchy relative to others. As such, Rancière holds that equality is not presupposed because it is ethically or rationally desirable, but because it is 'structurally necessary' (James, 2014: 113). Thus, in a dialectical turn, Rancière asserts the ontological dimension of this idea as 'the ultimate secret of every social order, the pure and simple equality of anyone with everyone' (Rancière, 1999: 79). As a presupposed equality also presupposes the capacity for the recognition of structural inequality (Honneth and Rancière, 2016), this view of equality thereby supplies potential for challenge and change by facilitating an appreciation of the contingency of hierarchies. Consequently, the 'equality' theorised by Rancière holds promise for recognising and resisting the contingency of what he terms the 'distribution of the sensible' (Lennon and Moore, 2019).

In essence, Rancière is here articulating the manner in which our experiences of the world are inevitably communal insofar as 'any world can only be experienced as such on the basis of a horizon of perception which is common to all those who inhabit that world' (James, 2014: 118). However, within such a common horizon, sensible experience is segregated according to the partition of sites and spaces that govern where one is situated within that world and the particular interpretive frameworks one uses to negotiate that world (Tanke, 2011). Hence, the 'distribution of the sensible' is 'the system of self-evident facts of sense perception that simultaneously discloses the existence of something in common and the delimitations that define the respective parts and positions within it' (Rancière, 2004b: 12). Rancière does not view the distribution of the sensible as open to easy reconfiguration. Rather, he believes that it is maintained by a 'police order' that reflects the hierarchical manner in which interpretation is structured. In an idiosyncratic re-description of a familiar term, Rancière's use of the word 'police' here does not refer to a body of people who maintain order (i.e. 'the cops') or any other repressive organ of the state. Rather, he is echoing an interpretation more closely 'identified by Foucault in seventeenth- and eighteenth-century writings as synonymous with the social order in its entirety' (Davis, 2013: 76). Hence, the 'police' denotes the prevailing distribution of the sensible[4] and how it is reinforced by an assumption that all entities are accounted for and participate within 'a whole as the whole with each part in its proper place' (Dikeç, 2005: 175). 'As such, the "police" is rather close to Foucault's notion of governmentality, the conduct of conduct, the mode of assigning location, relations and distributions or what Alain Badiou refers to as "the state of the situation"' (Swyngedouw, 2009: 606). While apparently similar to Mouffe's construal of 'hegemony,' the police order differs in its recognition. Thus, the adversarial politics of Mouffe's thesis facilitates the easier identification of a hegemonic order to be resisted (e.g. patriarchy, neoliberalism, scientism). However, the police order is a more closed arrangement of what is/can be perceived rather than an identification of what is currently dominant. As summarised by Chambers (2010: 63),

> This logic can be completed as follows: police determines not just the part that any part has in society; it also determines the intelligibility of any party at all. To have no place within the police order means to be unintelligible – not just marginalized within the system, but made invisible by the system. Police orders thereby distribute both roles and the lack of roles; they determine who counts and they decide that some do not count at all.

In this sense, the post-political dimension resides in the specifics of how the police order is normalised. The important turn for post-political thinkers is thus how Rancière's particular idea of 'equality' entails a capacity to identify the contingency of hierarchical distributions, thereby supplying potential for reflection on the distribution of the sensible and the possibility of resistance to the police order (Dikec, 2015). Rancière refers to the process of resisting and contesting the police order as 'politics' in what is clearly a different use of the term to that employed by Mouffe.

For Rancière, 'politics is an activity of reconfiguration of that which is given to the sensible' (Rancière, 2000: 115). As such, it revolves around 'what is seen and what can be said about it' and 'around the properties of space and the possibilities of time' (Rancière, 2004b: 13). In this sense, politics is about challenging the contingency of the police order. 'Disagreement in the proper political sense hence revolves around a conflict over the *distribution of the sensible*' (Metzger et al., 2015: 8 – emphasis in original). Rancière proposes that recognition of difference leads to a reconfiguration of the police order by introducing new identities and agents. This is achieved through a process of 'subjectification'. He employs this term inversely to the more familiar form of subjectification discussed by Foucault where power relations create subjects. For Rancière, the notion of subjectification entails an ontological presupposition of equality that facilitates an appreciation of the police order's contingency and may prompt a striving for recognition (Lennon and Moore, 2019). As he explains,

> By subjectification I mean the production through a series of actions of a body and a capacity for enunciation not previously identifiable within a given field of experience, whose identification is thus part of the reconfiguration of the field of experience.
>
> *(Rancière, 1999: 35)*

Rancière believes that the identity of that which emerges is neither the source of the subjectification activity nor its outcome. Rather, 'it emerges alongside the ongoing activity, feeding and being fed by it' (May, 2010: 79). Hence, subjectification is a complex, emergent yet discernible process in which the subject in question may serve as 'an operator that connects and disconnects different areas, regions, identities, functions, and capacities existing in the configuration of a given experience' (Rancière, 1999: 40). This leads Van Wymeersch et al. (2019: 363) to conclude,

> Contrary to Mouffe, Rancière (1999) does not define a political subject as a group that becomes aware of itself, finds its voice, forms a counter-hegemonic bloc and imposes its weight on society (p. 40). Nor does he believe – as Habermas does – that the political is situated in subjects performing communicative reasoning in order to reach consensus on the common good.

Van Puymbroeck and Oosterlynck (2014) conjecture that this 'relational approach' (Metzger, 2018: 183) to understanding the subject and their positioning within a contingent, yet seemingly 'given' order of intelligibility, permits a convincing analysis of the multiple ways that a political impasse can be addressed by exposing the contingency of any current configuration and revealing possible trajectories for change. Indeed, a fast-growing reservoir of work employing Rancière's ideas has emerged in recent years in urban studies and planning (Bassett, 2014; Dikeç, 2005; Gualini et al., 2015; Metzger, 2018). Much of this has occurred at the macroscale with a focus on valuable but broad commentaries regarding neoliberalism and post-politics (Davidson and Iveson, 2015; Dikeç and Swyngedouw, 2017; Metzger et al., 2015; Wilson and Swyngedouw, 2014), although a nascent line of research is attending to the micropolitics of participatory planning and land-use designation (Lennon and Moore, 2019; Van Wymeersch et al., 2019). These studies generally position planning as a 'police order' that instantiates a form of sensibility which results in the suppression of alternative ways of seeing and doing. Hence, planning appears as a foil to the struggle of individuals and groups for recognition. Furthermore, the weight of the state and the neoliberal ideologies it sustains render precarious any successful effort at subjectification by potentially checking the duration and scope of modifications to intelligibility. For as noted by Rancière (2004a: 7), subjects 'are always on the verge of disappearing, either through simply fading away, or more often than not, through their re-incorporation.' This view, coupled with Rancière's contention that it is the agency of the subject that results in subjectification, leaves a constructive role for planning – and by extension the planner – somewhat absent in this work. Indeed, the prevalence of Rancièrian research that examines macro-level issues of ideological mobilisation usually renders the planner quite opaque while implicitly portraying them as instantiating the police order through activities that give expression to neoliberalism in their practice.

This obscure but tacitly profiled complicit planner reflects a broader wariness of the 'common good' as a concept. Indeed, the post-political critique emerged as a response to what was perceived as the displacement of political difference and is 'clearly linked to the rise in consensus-based, 'Third Way' thinking in politics across Europe and the USA in the 1990s' (Allmendinger, 2017: 201). Despite differences in approach, post-political theorists are thereby united by a disquiet on the foreclosure of debate via consensus-seeking planning processes. In this sense, 'there exists a strong agreement among analysts of post-politics that this is a phenomenon that is devious and destructive and which therefore needs to be uncovered, countered and opposed' (Metzger, 2018: 188). Given this suspicion of consensus and the role of the state in advancing neoliberal agendas that are largely viewed as damaging to democracy, post-political thinkers hold appeals to the 'common good' as attempts to quell debate, foreclose questioning and advance programmes that suppress difference. In this, post-political theorists share with Foucauldian planning scholars a distrust of endeavours to specify what the common good is or how it can be identified. However, in striving to prevent the closure of politics, whether conceived in a Mouffian or Rancièrian sense, post-political planning theory implies a conception of democracy characterised by interpretations of liberty and equality, which it is vital to maintain for the operation of an open pluralistic society, a constitution of society that is worth protecting to prevent degeneration into violence (Mouffe) or injurious anonymity and quiescent repression (Rancière). Hence, there is a paradox at the heart of post-political theory: a common good cannot and should not be specified, yet there should be a commitment to pluralism, liberty and equality as attributes that when synergistically operative result in the shared (common) good of democracy-saving 'conflictual consensus' (Mouffe) and effective 'subjectification' (Rancière) that can 'de-articulate hegemonies' (Mouffe) and reconstitute the 'police order' (Rancière).

Forgetting and Forsaking

While philosophical differences have always characterised planning, the diversity of thought is greater now than ever. This led Innes and Booher (2015: 196) to conclude that, 'Today planning theory seems to have become a set of dividing discourses. People talk past one another . . . Theorists belong to discourse communities which employ different languages and methods towards different ends.' Indeed, as traced through the dialectic of the present and the previous chapters, planning theories often emerge as consciously contoured counterpoints to each other, furnishing different ontologies that delineate different problems and offer different solutions. Distinguishing unity in this vociferous field of contending voices is not an easy task. However, by focusing on the question of 'why' do planning, the challenge is made possible through sensitivity to the work of Paul Ricoeur. As a philosopher in the arcane field of literary hermeneutics, Ricoeur's oeuvre may not at first appear an obvious location to look for a way of conceiving unity in planning. Nevertheless, his work on 'character' supplies guidance for thinking about how the identity of planning as a discipline that has changed over time and has multiple contemporary voices remains cohesive despite theoretical disagreements, disparate foci and dissimilar methods. Of particular note is Ricoeur's discerning a distinction between the meanings of 'sameness' in the concept of identity.[5] Referencing Latin, he extracts two meanings of identity: the first is *idem*, which equates to 'being the same'; the second is *ipse*, which preserved in the English word 'ipseity,' means 'selfhood' (Ricoeur, 1995). The essential difference is that idem means identicalness while ipse means self-constancy. Therefore, whereas idem denotes that something is fixed, ipse conveys a sense of change that is nevertheless unified through an identity that facilitates difference. In this sense, the ways in which planning changes yet retains its sameness is ipse, while the constant sameness of using the designation 'planning' to describe a profession at any point in space and time or across varying activities is idem. According to Ricoeur, these two dimensions of identity fuse in the idea of 'character'. Importantly, Ricoeur holds that character is forged through a moral commitment kept over time in extended reflection, debate and engagement. Seen in the context of planning, this can be understood to map onto debates within the discipline and engagement with others not directly identified with it (e.g. the broader public, politicians, allied professionals and scholars). Hence, while different elements of constancy and change are perceptible in planning's evolution, different perspectives on what planning is and should do retain a discernible identification with the discipline's underlying moral commitment to 'do good,' albeit what it means to do good may be variously conceived and implicitly or explicitly expressed. Indeed, as noted by Wachs (2016: 464),

> planning has historically been about shaping our shared built environment, but over time it has also come to be about forming our collective institutional and social environments . . . . Every collective or social decision is based in part on explicit or implied moral values, and it is inevitable that every act of planning is to some extent inspired by thought about morality. Programs addressing housing, air quality, mobility, and economic development all have complex technical content but are motivated ultimately by social concerns about achieving the right and the good.

In a similar vein, Campbell notes, 'In many respects to plan is to conceive of the future; a future, hopefully, rather better than the present but at least no worse' (2003: 461). Hence, in seeking to create a better (or no worse) future, planning is an inherently normative enterprise. Accordingly, what threads the various manifestations of planning together across different contexts – what gives planning its 'character' – is a commitment to making things better by advancing theory and practice in administering interactions with and within built and natural environments. If approached from this perspective, the expansive family of planning thought and action come to share the same space. This is because what justifies planning as an activity and what it has always sought and continues to seek is a 'good' or series of 'goods' that in some way is justifiable as a common benefit, even if as Gunder and Hillier (2009: 191) contend, such goods (multiculturalism, sustainability, etc.) are simply fantasies created to suture the yawning gap between the desired and the experienced in sustaining the realities that planning creates for itself.

For some, doing good involves enhancing the science (systems thinking) and procedures (rational process thinking) of planning. However, gathering pace from the late 1980s, growing criticism was levelled against such views as operating on an outmoded form of unjustifiable exclusion that erodes confidence in the profession. Working on this basis, a new breed of theorists sought to do good by developing a theory that would foster the inclusion of impacted publics in decision-making. This was advanced through formulating innovative consensus-seeking approaches within the micro-dynamics of practice (communicative planning theory). Approaching and extending into the twenty-first century, theorist have built upon the platform of such thinking. Mining new philosophies and often in contradistinction to earlier programmes, these scholars have sought to reveal the insidious role of power in planning (e.g. Foucauldian planning theories) as a means of sensitising theorists and practitioners to its indelible presence. This has been undertaken as a way of doing good by revealing how power is negotiated, as well as how its intentional abuse or inadvertent effects can be identified and addressed. Recent progress in scholarship has endeavoured to do good by turning attention to the ways in which the consensus-seeking aspirations of some planning theories can undermine meaningful participation in planning through foreclosing fundamental questions on the trajectory of planning objectives (e.g Mouffian post-political theory) or exclude viewpoints, identities and dispositions that are non-aligned to prevailing modes of understanding within planning (e.g. Rancièrian post-political theory). Other perspectives are also circulating in planning theory. These include works informed by the thinking of Lacan, (Gunder, 2014) Deleuze (Hillier, 2018) and Latour (Rydin and Tate, 2016), as well as novel research on complexity theory (Chettiparamb, 2014) and innovative scholarship in institutional theory (Salet, 2018a).[6] This exciting research is taking planning theory in new directions. While generally more descriptive than prescriptive in their approaches, all seek to offer fresh ways of thinking about what planning is and how it operates. This can be used to forge better understandings of what ought to be done and why. As such, the 'good' that they provide is greater clarity. Hence, planning theory and practice of all hues are inherently motivated and profiled by an implicit or explicit appeal to a conception of a good justifiable beyond the bounds of the specific ideas adumbrated or actions undertaken. The good of this good is thereby its common benefit. To be a 'common good' in this sense is not to specify the particularities of a defined concept, be it justice, sustainability, recognition, clarity or any of the host of surrogates frequently deployed since the postmodern turn to diversity accelerated a decoupling of the 'common' from the 'good' in planning parlance. Nor is it to accentuate a utilitarian view that the preeminent 'good' should be determined by its aggregated commonness. Instead, it is to appreciate that the propellant which drives theorists and practitioners to search for new ways of thinking and doing is an enterprise shaped by its common aim to 'do better,' whatever this may mean.

It is therefore interesting to note that in the current trajectories pursued by the most influential streams of planning theory, the 'common good' concept is increasingly dissolved into the dispersed spaces between debates such that it now seems to provide little more than a hazy backdrop that is sometimes vaguely referenced, occasionally denounced but most frequently ignored. Thus, we theorise at a juncture where the discipline is driven by various interpretations of a common good, yet our understanding of what this may mean has rarely been so diluted by suspicions that carry us away from considering it as an issue of scholarly concern. Accordingly, only a few theorists now directly engage with this concept (Alexander, 2002b; Alfasi, 2009; Campbell and Marshall, 2002; Chettiparamb, 2016; Grant, 2005; Lennon, 2017; Moroni, 2004, 2018; Tait, 2016). Similarly, the moral dimensions of the 'planning subject' have been thinned out and largely overlooked as planning theory has evolved and expanded. Indeed, the planner of planning theory is now less a person with views, beliefs and commitments than a vaguely conceived scaffold to hang tasks resonant with the broader theoretical frame being specified. While hesitance for a more profound engagement with the moral planning subject may stem from uncomfortable associations of the planner with arrogance (Jacobs, 1961), ignorance (Young, 2000) or the inequities of capitalism (Harvey, 1973), the hollow planner of much planning scholarship means that we now possess an array of theories *in* planning and theories *of* planning but few theories of 'the planner'. As demonstrated in this and the preceding chapter, since the expert planner of 'rational process planning' and 'systems thinking' dissipated in a wave of post-positivist critique, a confused moral profiling of the planning subject seems to fall out of theory once the other pieces have been detailed. Hence, the deflating and de-centring of the planner has developed into an elision of the planning subject's moral dimensions.

And so, the forgetting of the moral planning subject parallels the forsaking of the common good. However, as the planner is deeply implicated in, if not the bearer of the variously understood implicit and explicit ideas of the common good that propel and mould the theories being advanced, the absence of attention to both is problematic for the discipline. This is because the ongoing superficial consideration or downright omission of the inherent enmeshing of both planner and the common good limits our capacity to debate on what planning is and should be, does and should do, by obscuring the very impulse for action and the bridge that connects theory with the world. Therefore, understanding how and why the planner reasons what they 'should' do to 'do better' is necessary to understand 'why' what is theorised as done or to be done matters if it is not to become meaningless. Consequently, as academics and practitioners, there is an imperative to redress the dearth of attention given to our appreciation of the common good, the planning subject and how these interlace. This is required if we are to make sense of how the moral politics of a polity's planning activities come to be organised around concepts directed at delivering a better (or no worse) future. As subsequent chapters demonstrate, this involves rethinking how the common good and the planning subject are understood.

Notes

1 Paul Ricoeur (1970) famously labelled the triumvirate of Marx, Nietzsche and Freud "the school of suspicion" in a moniker meant to denote those thinkers who regarded with suspicion our presumptions, learned understandings and interpretations of experience.
2 Whereas Slavoj Žižek is also associated with this movement and has been influential in the fields of sociology and geography among others, his work has yet to have a significant

impact on planning theory beyond the Lacanian critical social theory of Gunder and Hillier (2009). This is perhaps due to Žižek's advocacy of 'revolutionary change' and a preoccupation with broad structural debates, such as capitalism versus socialism (however see Stavrakakis, 2011).
3 See Hillier (2003) for a lucid explanation of the influence of Lacanian thought on Mouffe's theorising.
4 Although Rancière contends that his concept of the police is 'non-pejorative' (Rancière, 1999: 29) and that some form of police is inevitable, he does accede that 'there is a worse and a better police' (Rancière, 1999: 30).
5 It is noted that Ricoeur's target is the concept of 'selfhood' and not 'planning theory'. Nevertheless, in showing how the sense of selfhood is constituted by 'the dialectic of concordance and discordance developed by the emplotment of action,' he reveals how a unifying 'character' emerges from a totality which is itself 'singular and distinguished from all others' (Ricoeur, 1995: 147).
6 More accurately speaking, Salet's work seeks to reconcile institutional theory with pragmatism. He advances a well-argued thesis that 'planning needs both the dialectic of the practical and the institutional judgment, and we have to investigate their mutual contacts' (Salet, 2018a: 63).

PART 2

> The strength of any intellectual project lies in its diversity and disagreements, not in its consensus.
>
> *(Beauregard, 2020: 118)*

Part 1 dispelled any assumptions that the concept of the common good is unanimously valued across the spectrum of planning theories. Yet, it was argued that an implicit understanding of a diversely conceived sense of the common good nonetheless underpins the prominent theoretical perspectives that frame debate in our discipline. While at first this may appear wholly procedural, deconstructive or oppositional, a deeper analysis reveals a substantive commitment in each perspective to a good shared in common, even if this is to resist commonness. Thus, Part 1 concluded that despite its enduring relevance, the concept of the common good remains variously ignored, misunderstood or rejected in much planning theory. Similarly, it was shown that our understanding of the planning subject has been neglected by theory. It was argued that this impedes our understanding of what propels planning theory and practice. Hence, building on the review of Part 1, the chapters of Part 2 set out to redress this theoretical deficit by profiling the relationship between planning, the planner and the common good. Specifically, Chapter 3 mobilises elements of thinking from the moral philosophy of Alasdair MacIntyre to reconceive the inherent relationship between the common good and planning. This is then used in Chapter 4 as a platform from which to develop a nuanced understanding of the planner as a moral subject. This is achieved by extending and integrating the insights from MacIntyre's philosophy with those of Charles Taylor on the philosophical anthropology of the moral self. Taken together, these two chapters furnish the primary theoretical contribution of this book. In combination, they argue that there is an inherently co-constitutive and dialogical relationship between planning, the planner and the common good. Attempting to understand any one of these without due consideration of how it relates to the others results in the deficiencies of an incomplete picture wherein only a partial understanding is painted. Chapter 5 then illustrates such abstractions by grounding them in case study material.

3
PLANNING AND THE COMMON GOOD

The Persistence of Ignorance

When endeavouring to conceive planning's common good or its many synonyms (justice, sustainability, etc.), both theorists and practitioners alike frequently reach for seemingly neutral modes of reasoning and the impartiality of the standards these beget. Hence, much theory is infused with the deontological musings of Rawls and Habermas, while much practice is influenced by contemporary manifestations of Mill's utilitarianism (Campbell and Marshall, 2002). The appearance of detachment afforded by such reasoning lends planning practitioners a sense of expertise in bringing clarity to the turbidity of complex decision situations (Cowell and Lennon, 2014; Lennon, 2014). Indeed, a considerable volume of a planner's quotidian activities is justified on the basis of the discipline's capacity to identify the common good through the deployment of their training. It is on this basis that powers are delegated to practitioners apparently schooled in the craft of weighing multiple considerations in determining the best course of action. This is true of both policy formulation and the assessment of development proposals. Although decades of planning theory have problematised the simplistic view that planners know best on how to identify the common good and that planning is the practice of doing so, it nevertheless remains stubbornly embedded in planning practice, and consequently as a backdrop for much planning theory. Indeed, even in instances where agonistic planning theory has been identified as manifest, the planner (or planning inspector) is still tacitly understood to possess the expertise to adjudicate on what is the best decision outcome having carefully listened to the evidence mobilised by opposing parties when seeking to 'win' the case. This understanding is institutionalised in the very operation of planning in such instances (McClymont, 2011).

So, if planning cannot simply point to the common good it espouses or the means by which it can be identified beyond a limited set of decisions in specific

DOI: 10.4324/9781003155515-6

situations, how are we to understand what the common good of planning means? This is an oddly unanswered question for a discipline whose legitimacy is predicated on its role in advancing this ethereal concept. One solution is to reconceive the common good of planning not as something to be found 'out there' but as a moral endeavour advanced from 'within' a history of reasoning on what the common good means. In this context, the moral philosophy of Alasdair MacIntyre comes into view as furnishing a suite of conceptual machinery useful for teasing apart the complexities of this 'view from within'.

An Alternative Path

MacIntyre's philosophy has evolved from Protestant Marxism through analytical atheism onto Aristotelian-infused moral theory and more recently Thomist Catholicism. The breadth and depth of his work are matched only by the many twists, turns and revisions it has taken. Thus, the wholesale or uncritical adoption of his perspectives is not endorsed here, particularly as many views seem at odds with the liberal nation state characterising the modern western democracies that this book addresses. As such, the use made of his work is sparing in the context of the broader spectrum of those topics he has written on. Yet even here, issue is taken with how his theorising on moral reasoning has developed from his magnum opus *After Virtue* (MacIntyre, 1984) to a view more conversant with the work of Thomas Kuhn and Imre Lakatos in the philosophy of science, which it is felt are inappropriately 'mapped onto' the architecture of his later philosophy (MacIntyre, 1988). Hence, his oeuvre has been discriminately employed in what follows to help unlock difficulties in our conception of planning.

MacIntyre's mission is to diagnose what he sees as the malaise of contemporary moral thought and sketch the outline of how an alternative might be conceived. In this, his critique takes particular aim at the ideological pretensions of modern moral philosophy through exposing the lack of self-awareness of those who explicitly or implicitly advance an 'objective' approach to reasoning what 'the good' entails. He traces this condition back to the 'Enlightenment project' that promised to release humanity from the intellectual restrictions born of explanations framed through unscientific supposition. In essence, he conceives the Enlightenment project as resting on a desire to 'free' society from superstition and the perceived capriciousness of nature through the use of unencumbered reasoning such that 'the concept of freedom . . . constitutes the keystone of the whole structure of a system of pure reason' (Kant, (1788) 2012: 5.4). In this context, freedom is understood as freedom *from* the problems generated by defective thinking through promoting freedom *to* critically reflect on received wisdom (Lennon, 2017). Hence, it is conceived that freed from traditional modes of thought, the Enlightenment's 'objective logic' will facilitate both mastery of society (Hobbes) and mastery over nature (Bacon). Influenced by growing faith in the emerging scientific method as the arbiter of truth, perspectives on how to best reason about social and moral issues leaked out from the natural sciences to wash away the suspicions of tradition and bathe social and moral thinking in the new light of a unifying objective rationality. Given force by Enlightenment luminaries such as Descartes, Kant and Mill, this new rationality is founded on 'a rejection of metaphysical speculation and traditional moral authority, and in the belief that substantive human rationality is of only one kind, so that ethics may be advanced by rational argumentation that ignores anthropological and sociological factors' (Lutz, 2009: 53). According to MacIntyre,[1] such assumptions are metaphysical in nature and have influenced much moral thinking regarding the possibility of locating a unitary concept of moral rationality that tacitly mirrors the clarity brought by the natural sciences to our understanding of the physical world. Thus, in summarising the root of such thinking in the work of Enlightenment philosophers, he asserts,

> It was part of what I am calling their unitary conception of rationality and of the rational mind that they took it for granted not only that all rational persons conceptualize data in one and the same way and that therefore any attentive and honest observer, unblinded and undistracted by the prejudices of prior commitment to belief would report the same data, same facts, but also that it is the data thus reported and characterized which provide enquiry with its subject matter.
>
> *(MacIntyre, 1990: 16–17)*

In contrast, MacIntyre repudiates the assumption that it is possible to locate a 'view from nowhere' (Nagel, 1986) on moral matters. In doing so, he takes as his first target those forms of deontological reasoning which hold that moral action must be legislated for and that the condition under which all such legislation must fall is that of impartial rational justifiability. His second target is utilitarianism, which reasons a calculus for the objective determination of proper action. As summarised by Pinkard (2003: 187), MacIntyre believes that both these perspectives are fundamentally flawed in failing to account for the fact that

> it is only in light of our conception of specific types of goods that we form notions of having reasons and reasoning correctly, and those conceptions of goods themselves have a history and are not simply 'seen' by us or follow directly from some very general conception of formal reasoning.

This is because MacIntyre contends that a moral philosophy characteristically necessitates a sociology. Hence, he concludes that

> every moral philosophy offers explicitly or implicitly at least a partial conceptual analysis of the relationship of an agent to his or her reasons, motives, intentions and actions, and in so doing generally presupposes some claim that these concepts are embodied or at least can be in the real social world.
>
> *(MacIntyre, 1984: 23)*

It is on this basis that MacIntyre is sympathetic to the anti-enlightenment position advanced by Nietzsche, which he argues

> depends upon the truth of one central thesis: that all rational vindications of morality manifestly fail and that *therefore* belief in the tenets of morality needs to be explained in terms of a set of rationalizations which conceal the fundamentally non-rational phenomena of the will.
>
> *(p. 117)*

However, MacIntyre finds Nietzsche's perspective problematic on the grounds of the genealogical method he advances. Whereas Enlightenment thinkers presumed that objective rationality could be legitimated through deontological definition or via the neutral machinery of a moral calculus, Nietzsche's genealogical method sought to expose the contingency of concepts such as 'truth' by treating them as intellectual artifacts. Yet rather than leading him to consider the potentially positive contextual validity of such ideas, Nietzsche's genealogical method endeavoured to demonstrate the pernicious influence they exert. Accordingly, MacIntyre notes,

> Nietzsche, as a genealogist, takes there to be a multiplicity of perspectives within each of which truth-from-a-point-of-view may be asserted, but no truth-as-such, an empty notion, about the world, an equally empty notion. There are no rules of rationality as such to be appealed to, there are rather strategies of insight and strategies of subversion.
>
> *(MacIntyre, 1990: 42)*

According to MacIntyre, the problem with this view is that it undermines itself through immanent critique. This is consequent on how the Nietzschean genealogy that was proclaimed as 'the liberation of humanity from the deceptions of metaphysics and academic dogmatism also turns out to be self-deceptive' (Lutz, 2009: 55) because it is grounded on metaphysical presuppositions of timeless rational standards rather than escaping them. MacIntyre thus contends,

> In making his or her sequence of strategies of masking and unmasking intelligible to him or herself, the genealogist has to ascribe to the genealogical self a continuity of deliberate purpose and a commitment to that purpose which can only be ascribed to a self not to be dissolved into masks and moments, a self which cannot be conceived as more and other than its disguises and concealments and negotiations, a self which just insofar as it can adopt alternative perspectives is itself not perspectival, but persistent and substantial.
>
> *(MacIntyre, 1990: 54)*

Therefore, MacIntyre holds that Nietzsche's genealogical method subverts itself by tacitly resting on certain metaphysical suppositions regarding the prospective truth of a stable subject to ostensibly advance anti-metaphysical arguments designed to de-centre assumptions with respect to the possibility of truth and stability in the constitution of the subject. In this sense, it advances a claim to the truth that there is no truth, which undermines the very truth it asserts! MacIntyre believes that this issue is compounded by the intellectual heirs of Nietzsche, who have developed an inherently confused research programme. As summarised by Lutz (2009: 56), 'If the goal is to discover the truth, then genealogy has set out to do what it holds to be impossible. If the goal is something else, then the genealogical project, considered as a research program, is untenable.' Fittingly, MacIntyre finds himself uncomfortable with 'the liquidation of the self into a set of demarcated areas of role playing' (MacIntyre, 2007: 205) by existentialist philosophers, and the rejection of appeals to moral truths by postmodernists who perpetuate a research programme that ironically reverts to academic dogma on the truth about truth: a form of the very dogma it sought to overturn. It is in this context that MacIntyre (1990: 53) advances the example of Foucault to illustrate what he perceives as the irony in how implicit claims to truth and intellectual deference are cultivated by those who claim to undermine the justifications for these very attributes.

Therefore, it is between the hubris of moral philosophies influenced by the Enlightenment's desire for an objective ethics and the nihilism of Nietzsche's intellectual heirs that MacIntyre proposes an alternative route through the work of Aristotle. In doing so, he argues that Aristotle's teleological philosophy of moral reflection was inappropriately rejected as collateral damage to his discredited metaphysical biology. MacIntyre believes that the Enlightenment's early protagonists were correct to dismiss Aristotle's metaphysical biology in favour of the non-teleological mechanistic natural science they were developing. However, 'their mistake was to throw the baby of teleological ethics out with the murky bathwater of teleological metaphysics' (Knight, 2007: 134). Noting Nietzsche's undermining of the Enlightenment project, MacIntyre thereby sees some alignment between his position and that of the genealogists. In this sense, he argues,

> It was because a moral tradition of which Aristotle's thought was the intellectual core was repudiated during the transitions of the fifteenth to seventeenth centuries that the Enlightenment project of discovering new rational secular foundations for morality had to be undertaken. And it was because that project failed, because the views advanced by its most intellectually powerful protagonists, and more especially by Kant, could not be sustained in the face of rational criticism that Nietzsche and all his existentialist and emotivist[2] successors where able to mount their apparently successful critique of all previous morality. Hence, the defensibility of the Nietzschean position turns in the end on the answer to the question: was it right in the first place to reject Aristotle? For if Aristotle's position in ethics and politics – or something like it – could be sustained, the whole Nietzschean enterprise would be pointless.
>
> *(MacIntyre, 1984: 117)*

It is important to note here that 'in asking us to choose between Aristotle and Nietzsche, MacIntyre does not intend us to focus narrowly on the writings of these two philosophers. He treats each as types' (Solomon, 2003: 136). Nonetheless, given the aforementioned reservations MacIntyre holds regarding the genealogical methods born of Nietzschean philosophy, he understandably rests on the hope of rejuvenating a feasible Aristotelian perspective for moral philosophy. Hence, what MacIntyre recommends is a denial of the Enlightenment mindset 'that viable moral theory can find a basis for its enquiries or its conclusions independent of and external to moral practice' while rejecting the Nietzschean contention that thereby 'all moral theory is an illegitimate enterprise, one condemned to distortion and illusion' (p. 145). In charting his course between the Charybdis of the Enlightenment project and the Scylla of Nietzschean nihilism, MacIntyre proposes that 'the context within which theoretical moral enquiry alone has point and purpose is then that provided by the activities of some particular community' (p. 146). He thus proposes,

> What the Enlightenment made us for the most part blind to and what we now need to recover is, so I shall argue, a conception of rational enquiry as embedded in a tradition, a conception according to which the standards of rational justification themselves emerge from and are part of a history.
> *(MacIntyre, 1988: 7)*

Accordingly, central to MacIntyre's vision and key to understanding his attempt to reinvigorate Aristotelian moral philosophy is his conception of a 'tradition'.

Planning as Tradition

For MacIntyre, 'a tradition is constituted by a set of practices and is a mode of understanding their importance and worth; it is the medium by which such practices are shaped and transmitted across generations' (Mulhall and Swift, 1996: 90). In this sense, a tradition encompasses the debates on the modes of thinking, acting and evaluating in different fields of endeavour, such as aesthetics (e.g. painting, music, literature), occupations (e.g. carpentry, farming, teaching) and moral reasoning (Humanism, Catholicism, Confucianism). Although MacIntyre is silent on the difference between 'culture' and 'tradition,' it can be surmised from his work that a 'tradition' has a narrower focus and definition than a 'culture'. Specifically, a 'tradition' is focused on debates regarding the ways of thinking, doing and evaluating within a particular field of activity (moral, political, aesthetic), while a 'culture' signifies a much broader and more amorphous amalgam of numerous traditions. In essence, MacIntyre's theory 'embodies his criticism and rejection of the Enlightenment project of justifying morality, as well as his criticism and rejection of the conclusions of postmodern efforts to explain morality away' (Lutz, 2009: 33). For MacIntyre, a tradition embodies an explicitly or implicitly evaluative narrative account of reality that makes the world intelligible. This can be witnessed for example in the context of planning theory by such work as Foucauldian-inspired descriptions of 'the reality' of epistemic privilege and marginalisation (Boyer, 1983; Flyvbjerg, 1998; Huxley, 2006; Yiftachel, 1998) in the constitution, dissemination and institutionalisation of rationalities that allocates power and resources to some at the expense of others. However, as discussed in Part 1, numerous contending explanatory narratives can concurrently circulate among a community – in this case planning theorists – such that the skirmishes between traditions (e.g. collaborative planning vs agonistic planning) are not simply about the conclusions of the evaluative narratives provided but also concern the premises of those arguments. As such, reasoning on how one identifies what matters, on what is 'substantive' relative to a shared way of rendering the world morally intelligible, involves thinking and acting on a set of standards and beliefs that profile one's interpretation of reality. Here, what can be called 'reason' thereby 'names a set of social practices that involve the asking for and giving of reasons, the evaluation of those reasons and the asking for and giving of such, and, importantly, the evaluation of the good' (Nicholas, 2012: x). Hence,

> according to MacIntyre's account, rationality is inseparable from tradition, because substantive rationality, whether speculative and practical, is a kind of practical art. Like every practical art, substantive rationality develops within a community of practitioners who measure it by common standards. Rationality is not something separate or even separable from traditions through which traditions may be judged: rather, it is something arising from traditions themselves, and bound up with traditions.
>
> *(Lutz, 2009: 57)*

Consequently, at the heart of MacIntyre's thesis is the view that rationality is tradition-constituted and tradition-constitutive. This gives a rational argument on those substantive issues held dear by a tradition an historical dimension.

At this juncture, critics would likely position on the 'communitarian' end of the theoretical spectrum the argument that the substantive goods we observe are shaped by those communities we interact with in our professional and personal lives. While this is a rubric beneath which a collection of often contending perspectives are grouped, all nonetheless share an interest in the role played by one's social environment in affecting one's outlook. Some have accused such communitarian perspectives, and MacIntyre in particular, of holding a static view of 'the community' wherein the political dynamics that initiate change from within a community are overlooked by a mislaid focus on shared, rather than contested frameworks (Frazer and Lacey, 1993). This criticism assumes that MacIntyre simply eschews the politics of moral deliberation within a tradition. As he explains,

> We are apt to be misled here by the ideological uses to which the concept of a tradition has been put by conservative political theorists. Characteristically such theorists have followed Burke in contrasting tradition with reason and

> the stability of tradition with conflict. Both obfuscate. For all reasoning takes place within the context of some traditional mode of thought, transcending through criticism and invention the limitations of what had hitherto been reasoned in that tradition; this is true of modern physics as of medieval logic. Moreover, when a tradition is in good order it is always partially constituted by an argument about the goods the pursuit of which gives to that tradition its particular point and purpose . . . . Traditions, when vital, embody continuities of conflict. Indeed when a tradition becomes Burkean, it is always dying or already dead.
>
> *(MacIntyre, 1984: 222)*

Accordingly, on closer inspection, it becomes clear that what MacIntyre is doing is reinterpreting how change in a tradition occurs. For him, all moral enquiries commence not in a vacuum but from within the normative commitments of a tradition of moral reasoning located in space and time. In this way, he holds that

> objective moral criticism is not justified by any appeal to the objectivity of an ideal observer. Objective moral criticism is founded upon the best theories so far developed in response to the experiences and epistemological crises of a person or community who adheres to the experience of reality.
>
> *(Lutz, 2009: 81)*

Consequently, matters do not remain static. This is because those within a tradition continually encounter ethically complex real-world situations that force reflection upon moral reference points and internal contradictions such that the validity of the rationalities employed may be called into question (Porter, 2003). Therefore, some of the standards of evaluation dominating a particular tradition must occasionally change if that tradition is to endure. Thus, for MacIntyre, 'a living tradition, then, is an historically extended, socially embodied argument, and an argument precisely in part about the goods which constitute that tradition' (MacIntyre, 1984: 222).

Critics of this conception of tradition have levelled accusations of relativism against it, especially given MacIntyre's refutation of Nietzschean philosophy. Indeed, some have gone so far as to censure him for being 'a dangerous relativist (since he offers a radically pluralist concept of moral practices)' (Higgins, 2003: 279–280). For MacIntyre, this 'relativist challenge' emerges from the static understanding of tradition he counters in his dismissal of Burkean conservatism. Essentially, the accusation centres on a problem of circularity wherein 'adherents of a particular tradition judge a standard of rationality to be true because of their formation in their tradition, and they judge their tradition to be truthful because they hold that rationality' (Lutz, 2009: 69). It appears that MacIntyre does indeed conceive an approach to moral philosophy that is relative to an historically extended argument on substantive rationality (Miller, 1994; Mosteller, 2008). However, he qualifies this with the view that interrogating the contingency of the perspectives held by a tradition can help reform that tradition. In contending that there is no neutral position from which to judge the limitations of a tradition, he argues for the attentiveness of one's *situated* perspective when evaluating different traditions against the norms of one's own and in doing so allowing such alternate traditions to raise questions in one's own (Lutz, 2012: 178). Ultimately, criticisms levelled against MacIntyre as a relativist are therefore reasoned within a tradition of ethical thought inherited from the Enlightenment where universalism and relativism are seen as the sole and polarised options. As noted by Porter (2003: 46),

> the plausibility of both relativism and perspectivism derives from the fact that both reflect the inversion of the Enlightenment ideal of a universal valid standard of rationality and truth. Since this cannot be attained (and MacIntyre agrees it cannot), the only alternative, it is said, is some form of relativism or perspectivism.

Hence, MacIntyre believes that acknowledging one's perspectives as 'situated' can supply a more accurate reflection of how we operate in a contingent world. The conclusion MacIntyre draws thereby seeks to navigate a path between realism and relativism by steering a self-reflective route through 'relativity'. Here, when forming a position on a topic such as how to best understand planning,

> not only is 'the best theory so far' not 'the truth,' it is not even necessarily the best theory so far. Rather it is the theory that appears, to the best of our available rational recourses, to be the best so far.
>
> *(Lutz, 2009: 105)*

This is consequent on how our views on a topic are inherently situated within the historically extended evaluative narrative of the tradition from which we understand that topic. Thus, MacIntyre's path between realism and relativism charts a form of non-foundational moral realism that while furnishing a sense of solidity acknowledges that the reality it rests on can itself come to change as the substantive goods that are part of our reasoning evolve. In this way, he refuses to separate 'form' and 'content' into some 'rock-bottom distinction,' instead arguing that 'our conceptions of what counts as good reasoning is linked to our conception of those substantive goods about which we are reasoning' (Pinkard, 2003: 194).

For example, the way of reasoning on what constituted the common good in planning was vigorously contested by Jane Jacobs and others in New York during the 1960s and helped prompt a re-evaluation of where planning's common good lies, which in turn recalibrated reasoning on what may count as 'good planning' (Hirt and Zahm, 2012). Importantly, this was not achieved via an a priori determination of what ought to be done or an a priori prescription on how what ought to be done should be determined. Rather, it was realised by demonstrating how certain practices prevalent at the time undermined the good that planning could and should aim to achieve based on shared perceptions of planning's raison d'être as a force for the common good. This was an immanent critique that emerged from those involved in planning activity, be it as theorists questioning the direction of planning, practitioners affecting change or community members affected by such change.[3] As an internal dialectic both within and reflecting on planning, such debate sought to explore the 'ends' that planning ought to achieve instead of assuming that such 'ends' would simply materialise as a product of proper procedures ('means') or could be predetermined as an always applicable suite of objectives ('ends'). Accordingly, the self-reflective quality of debates on what planning should do meant that evaluating reasons was innately entwined with beliefs on what amounted to 'the good'. This implies that perceptions regarding the legitimate reasons for planning in the common good are co-constitutive with attempts to define the common good. This is because all judgements, including the standards for the evaluation of planning's ultimate objectives, are conducted within a tradition of reasoning what those objectives may be. In essence, it means that reasoning the common good is an exercise in substantive rationality. Therefore, affiliation with a tradition furnishes the conditions for a debate among parties who seek to determine what is choice worthy *and* what is a worthy choice.

We can thereby appreciate how Jacobs and her associates where able to stimulate a reinterpretation of 'the good' consequent on a process of substantively questioning the raison d'être of planning. This was achieved by demonstrating that the emergence of a modernist orthodoxy in planning was inhibiting realisation of the common good by destroying the spatial conditions and social structures needed for communities to flourish. This reasoning was used to show how such modernist perspectives were generating incoherence between the legitimacy of planning as a force for realising the common good and the reasons given for 'why,' 'what' and 'how' planning was been conducted. Thus, planning thought and action is neither predetermined by the dominant mores of contextual evaluation in a tradition of seeking the good nor is it fully free of them. Instead, the planner simply starts from a contextually contingent point within a tradition as a means of worldly engagement. From this position, the subject can reflect on, (re)interpret and reorientate that tradition should he or she identify logical inconsistencies within the modes of evaluation deployed by that tradition in different contexts when seeking to identify the good. Thus, for Jacobs and her associates, challenging the emerging orthodoxy of modernist planning ideas involved making judgements on what is 'good' and 'bad' planning relative to a framework supplied by intersubjectively constituted ways of reasoning on how to determine the common good. The immanent critique advanced by Jacobs was facilitated by a moral compass that provided direction when considering what 'ends' planning 'ought' to seek. Consequently, for Jacobs, determining what the common good amounted to in the context of 1960s New York involved reflectively engaging with her understanding of the tradition carried by planning as a self-reflective enterprise that seeks to create a better (or no worse) future. It was this engagement that helped reorientate and extend that tradition by elaborating how planning ought to conceive the common good.

So what then is the relationship between different theoretical and practice traditions *in* planning to the tradition *of* planning? This is best conceived as a relation between 'orders'. As concluded in Part 1, planning theory and practice are inherently driven and shaped by an inferred or overt appeal to a conception of a good justifiable beyond the specific ideas proclaimed or actions performed. The good of this good is thereby its shared benefit: its 'common good'. Observed on the theoretical terrain of how to understand planning activity, several first-order traditions thereby surface *in* planning. These are the numerous currents of planning theory that variously align, contest or ignore one another. Each is a tradition of substantive rationality based on premises that emphasise different foci, be that prediction (systems theory), inclusion (collaborative planning), power (Foucauldian theory), politics (agonism) or a host of other primary concerns. Each has its own historically extended internal debates and deliberations with other traditions *in* planning that help extend the explanatory depth and reach of its tradition. Yet taken together, they all seek to advance the tradition *of* planning as a normative endeavour to create a better (or no worse) future. Similarly, each of the professional activities undertaken by planners has its own substantive rationality concerned with issues ranging from the effectiveness of different approaches on placemaking to the proper application of policy regarding sustainability. Each such tradition is infused with historically extended internal deliberations that serve to enhance the functionality of its tradition in a meshing of moral and practical ambitions. Yet as with planning theory, all are orientated towards the creation of a better (or no worse) future. Thus, the history of each such first-order tradition, be it theoretical or practitioner focused, is constituted by and is constitutive of the substantive rationality of that tradition regarding what matters, how it does so and why? The tradition *of* planning is then a second-order tradition of thinking and acting that comprises these first-order traditions. The general schematic thereby repeats at a higher level but is both profiled by and profiles the first-order traditions it encompasses in seeking to advance the common good on which it is ultimately legitimated. Hence, in any decision situation, the determination of the correct course of action is moulded by the configuration of the constellation of substantive rationalities *in* planning that comprise the tradition *of* planning in a particular context. It is at such junctures that the 'practice' of seeking the common good is most easily discernible.

Planning as Practice

In conceiving the idea of a 'practice,' MacIntyre espouses the Aristotelian view that the notion of what it is right to do should be inseparable from the concept of what it is good to be. Aristotle viewed this as constituting the 'purpose' of one's being, both in terms of what one does and whom one is (Aristotle, 2014). However, MacIntyre does not suggest that the views of Aristotle can be simply contemporised to a world where the braiding of one's purpose with the metaphysical biology of classical antiquity is no longer tenable. Hence, he reinterprets Aristotelianism to speak of practices that are not predicated of individual beings, but as 'a kind of relation between humans as social actors' so that 'practices are functions of such social relations rather than of individual beings' (Knight, 2007: 146). MacIntyre extrapolates from this view an action-centred 'social teleology' (Lutz, 2012: 108), which he endeavours to contemporise by suggesting that the purpose of an activity is given expression through 'practice'. In this he maintains that a practice is defined intersubjectively within the historically situated form of reasoning on the good that is employed by those affiliated with a tradition (MacIntyre, 1984). Consequently, he offers a detailed definition of what a 'practice' is:

> By a 'practice' I am going to mean any coherent and complex form of socially established cooperative human activity through which goods internal to that form of activity are realized in the course of trying to achieve those standards of excellence which are appropriate to, and partially definitive of, that form of activity, with the result that human powers to achieve excellence, and human conceptions of the ends and goods involved, are systematically extended.
>
> *(MacIntyre, 1984: 187)*

MacIntyre supplies a selection of practice examples. Among others, these include biology, architecture and the work of the historian. Each of these practices is characterised by a level of social cooperation in their history, current state and future development. From MacIntyre's perspective, an activity entered into primarily for 'external goods,' such as monetary reward, does not amount to a practice, although receiving such a reward does not disqualify an activity from being considered a practice (MacIntyre, 1988: 35). As he explains,

> It is characteristic of what I have called external goods that when they are achieved they are always some individual's property or possession. Moreover, characteristically they are such that the more someone has of them, the less there is for other people. This is sometimes necessarily the case, as with power and fame, and sometimes the case by reason of contingent circumstances as with money. External goods are therefore characteristically objects of competition in which there must be losers as well as winners. Internal good are indeed the outcome of competition to excel, but it is characteristic of them that their achievement is a good for the whole community who participate in the practice.
>
> *(MacIntyre, 1984: 190)*

Thus, while external goods are attractive, they are not necessarily interlaced with the internal goods of a practice as MacIntyre sees it, since one may achieve external goods by means other than practice excellence, such as luck, deception or theft. Accordingly, 'it is an essential characteristic of every form of cooperative human activity that genuinely qualifies as a practice, that excellence in that activity requires attention to and perseverance in the pursuit of its internal goods' (Lutz, 2009: 97). MacIntrye's concept of a practice thereby has an intrinsically 'moral-political' (Schwandt, 2005: 330) character that is principally defined by a self-referential characterisation of purpose beyond the instrumentalist pursuit of rewards other than the satisfaction of advancing practice excellence. Hence, a subject's 'motivation' is central in defining what counts as a practice. In this context, Lutz (2012: 157) usefully notes,

> Practices have four distinctive characteristics: (1) People pursue the practice because they want to, because it addresses some need or desire they have by providing certain goods. (2) A practice has internal goods. There are things that can be gained only through participation in the practice, and it is the pursuit of these goods that leads to true excellence in the practice. . . . (3) A practice has standards of excellence that develop along with the practice. (4) The success of the practice depends on the moral character of its practitioners.

Therefore, 'what characterizes a practice, in contrast to a skill or technique, is its orientation towards intrinsic goods that can be attained only through the practice itself, and that require both skill and sensitivity to the aims of the practice in order to be realized' (Porter, 2003: 40). As explained by MacIntyre,

> a practice, in the sense intended, is never just a set of technical skills, even when directed towards some unified purpose and even if the exercise of those skills can on occasion be valued or enjoyed for their own sake. What is distinctive in a practice is in part the way in which conceptions of the relevant goods and ends which the technical skills serve – and every practice does require the exercise of technical skills – are transformed and enriched by these extensions of human powers and by that regard for its own internal goods which are partially definitive of each particular practice or type of practice.
> *(MacIntyre, 1984: 193)*

Hence, from a MacIntyrean position, a practice is the source of the standards in the substantive rationality of a tradition as 'there are no standards prior to practices, because standards arise organically from the practices themselves' (Lutz, 2009: 41). In this way, practices are the motor of traditions that maintain their dynamism. This is because

> practices never have a goal or goals fixed for all time . . . the goals themselves are transmuted by the history of the activity. It therefore turns out not to be accidental that every practice has its own history and a history which is more and other than that of the improvement of the relevant technical skills.
> *(MacIntyre, 1984: 194)*

As such, 'to consider the history of a practice is to consider the development of the standards of that practice along lines dictated by the practice itself, and made intelligible and possible only by the preceding events of that same process of development' (Lutz, 2009: 41).

Conway (1995: 88) has criticised MacIntyre's concept of a self-referential practice on the basis that he 'has provided no good reason for thinking that, where people engage in certain productive activities for the sake of income or profit, those activities are precluded thereby from being instances of practice in his sense.' Conway cites the work of the architect as an example. However, what Conway misses is that MacIntyre views practices as moral enterprises whose evaluation should not be referenced to their worth in the marketplace, but rather must be judged against the standards of excellence carried by a tradition. To use Conway's own example, it is not financial gain for a subject's endeavours that matter, as it is possible to design a poorly functioning building and still receive significant financial reward. Instead, what matters is that one sincerely commits oneself to designing a building to the best of one's ability within the context of the evaluative standards of architecture's practice tradition (Lennon, 2017). For,

> to enter into a practice is to enter into a relationship not only with its contemporary practitioners, but also with those who have preceded us in the practice, particularly those whose achievements extended the reach of the practice to its present point. It is thus the achievement, and a fortiori the authority, of a tradition which I then confront and from which I have to learn.
>
> *(MacIntyre, 1984: 194)*

Accordingly, the different 'first-order traditions' of planning theory, such as collaborative planning or those critical approaches informed by Foucauldian perspectives, each depend on practice excellence on the part of their theorists. This involves sincerity of engagement with the past, present and potential future of their tradition in ways that reflect upon and extend that tradition's explanatory and evaluative depth when interpreting the world encountered. A parallel process exists in the 'first-order traditions' of planning practitioners, which to various degrees are interlaced with the arguments of planning theory. Moreover, corresponding to the relationship between the first and second-order traditions of planning, there is an equivalent association between the first- and second-order practices of planning. To unpack this correlation, consider the reflection on planning offered by Tewdwr-Jones (2012: 1):

> Planning as an activity that attempts to manage spatial change would not exist in any meaningful way if it was not for contention over the future use and development of land. Spatial planning is owned by everyone who has a vested interest in the land and what happens to it.

By conceiving planning in this way, it may first seem improbable that planning resembles the concept of a practice advanced by MacIntyre as it is frequently focused on mediating between competing issues concerning spatial governance drawn from a variety of practices rather than engaging with the internal development of practices per se. For example, planning may be more concerned with mediating between competing interests in nature conservation or the protection of historically significant architecture than with developing the science of restoration ecology or formulating new construction methods for built heritage conservation (both of which are practices within their respective traditions). Moreover, it may at first appear uncertain that there are discernible 'internal goods' of planning that can be orientated relative to identifiable 'standards of excellence'. Indeed, the inherently political character of planning would seem to render doubtful that it conforms to MacIntyre's understanding of a practice as a 'coherent and complex form of socially established cooperative human activity.' However, should planning be reconceived as the practice of arbitrating between the various competing issues that manifest in deciding how best to order social and social-ecological interactions in space, the idea of planning as a practice in the MacIntyrean sense begins to emerge. What is key in this context is how the activity of planning relates to the ordering of those concerns advanced by the established traditions of others, as well as by one's own (e.g. urban design, community development, transport planning, ecology, archaeology, landscape architecture). Murphy provides direction here when inferring that the very activity of ordering competing demands constitutes a practice:

> So consider the question that arises in communities of any complexity: how are the practices, and the goods internal to these practices, to be ordered in this community? This is a practical question: it is not for the sake of speculation but for action. There is a range of excellences that are necessary for answering this question well, and there is a range of capacities that are developed through successive attempts to answer these questions in common. An adequate explication of these excellences and the developed capacities, and of the worthwhile activity engaged in by those attempting to answer this question, cannot be offered except in terms of the activity itself. There are internal goods to the activity of attempting to answer questions about how the practices in a community's life are to be ordered. This activity is, therefore, a practice.
>
> *(Murphy, 2003: 163)*

Hence, planning may be considered a complex 'second-order' practice characterised by social cooperation among a community of practitioners wherein the tradition's standards of excellence profile the evaluative deliberation about, mediation between, sorting of, and adjudicating on 'first-order' practices. This ordering activity is ultimately orientated towards advancing the common good of creating a better (or no worse) future. The goods internal to planning are the benefits of being involved in and feeling a sense of worth that one is serving the common good in 'trying to achieve those standards of excellence which are appropriate to, and partially definitive of, that form of activity' (MacIntyre, 1984: 187). In the context of planning, the standards of excellence against which this practice is judged are defined by 'everyone who has a vested interest in the land and what happens to it.' For this reason, the standards of excellence against which planning practice is evaluated are dynamic rather than static as the performance of planning is referenced to the ongoing evolution of perceptions on how successful planning activity is in 'managing spatial change'. Thus, the standards of excellence against which planning practice is assessed are the aspirations for proper conduct and desired outcomes relative to the history of deliberation and current debates on how to advance the common good (or its synonyms) in the context in which the decision situation arises. The standards of excellence against which this practice is evaluated are thereby those inherited, (re)interpreted and transmitted by the tradition of reasoning about 'the good' in a decision situation within a particular spatiotemporal context.

Importantly, this conception of planning as a second-order practice means that it may be tradition-transformative rather than statically repetitious. This is because planning operates in a world of 'wicked problems' (Rittel and Webber, 1973). The existence of such a world demands that the subject engaged with planning is unavoidably confronted by debatable moral considerations in determining how best to identify the common good. This position means that judgement on how to advance the common good cannot be inflexibly predetermined or simply produced via a universally applicable procedure. Rather, such a subject must remain attentive to the way judgement on how best to promote the common good may emerge through a dialogue between tradition-informed perceptions of 'the good' and the moral complexity of those contexts encountered. Accordingly, from a MacIntyrean perspective, the standards of excellence against which planning is evaluated is context dependent and profiled relevant to the forms of substantive rationality operative within a community of practitioners on how adequate planning is in handling the issues of moral complexity encountered in a decision situation when seeking to advance the common good.

Planning and the Common Good

What MacIntyre offers is a teleological account of practice. In this, the precepts that are internal to practices are (re)interpreted in the thought and action of practitioners, are secondary to the shared aims internal of the practice that give it 'point and purpose'. What defines a practice is then 'the intentionality – and indeed, the social reality – of what he calls a shared "telos"' (Knight, 2007: 223). Thoughts and actions are thereby rendered intelligible by a narrative unity of intentionality, which is central to the identity (*ipse*) of planning as an endeavour seeking to advance the common good. This narrative unity in the tradition *of* planning forms the precondition for the trajectory of theory and activities *in* planning. Hence, this narrative is teleological in that it presupposes the possibility of evaluating successes and failures in terms of an orientation towards a good that transcends the occasions of such successes and failures. For as noted by MacIntyre (1984: 223), 'it is rather the case that an adequate sense of tradition manifests itself in the grasp of those future possibilities which the past has made available to the present.' He suggests that this self-reflective orientation of practice to reasoning the good via the substantive rationality of an evolving tradition can be understood as a 'quest'. As he explains,

> a quest is not at all that of a search for something already adequately characterized, as miners search for gold and geologists for oil. It is in the course of a quest and only through encountering and coping with the various particular harms, dangers, temptations and distractions which provide any quest with episodes and incidents that the goal of the quest is finally understood. A quest is always an education both as to the character of that which is sought and in self-knowledge.
>
> *(p. 219)*

Conceiving the tradition of planning as a quest for the good of the common good with an incomplete yet orienting telos, depicts it as a communal endeavour threading together a past, present and future, whose commitment transcends the particularities of specific activities and theories at any juncture in the discipline's evolution. In this way, the understandings that guide our characterisation of the common good represent moments in the development of the substantive rationality in the tradition *of* planning. In the course of this quest, it can thereby be expected that those understandings will evolve, but the quest cannot begin unless there is first an end to pursue. Otherwise, the theories and activities that are constitutive of and constituted by the tradition *of* planning run the danger of being rendered arbitrary. Hence, what characterises the tradition and practice *of* planning – its *ipse* identity – is a teleological orientation in an unfolding quest for the common good, even though what is seen to constitute this good may be contested at any point in this quest.

Notes

1 I follow here the tracing of MacIntyre's thought provided by Lutz (2009).
2 Echoing earlier criticism by Anscombe (1958), MacIntyre believes that the result of detaching the moral subject from the forms of shared moral reasoning that anchor one's outlook is to generate a confusing semblance of moral deliberation without an actual commitment to values. The outcome of this process of moral individuation is the emergence of what he terms 'emotivism,' a term he borrows but significantly reworks from that originally advanced by Stevenson (1963). For MacIntyre, an emotivist discourse is one where ethical language and moral reasons are advanced to justify action without recourse to a shared commitment to reasoning on the morals that are advanced. Hence, for a morally individuated subject to forward moral arguments is to create an obstacle to true moral deliberation by wilfully or unknowingly seeking to manipulate others for his or her own purposes. Consequently, appealing to the formality of rules and procedures (deontological justifications) or processes (utilitarian moral calculi) as ways to produce 'the good' is to confuse 'means' with 'ends'. Here, interlocutors engaged in 'rational' deliberation become 'means' to each other for the fulfilment of their subjective preferences rather than co-investigators endeavouring to realise a shared conception of 'the good'. Thus, it is not that 'appeal' to moral arguments concerning right and wrong or good and bad have been suspended, as it is plain that these continue to suffuse debate. Instead, what is absent in emotivist discourse is awareness that moral individuation entails the lack of a shared sense of how to identify and converse about what matters and why. MacIntyre believes

that interlocutors in such a discourse thereby risk eclipsing recognition of how they may inadvertently transform ostensibly moral deliberation into manipulative manoeuvring through the dissociation of terms such as 'good' and 'bad' from shared premises of moral reasoning that give these terms a common sense of meaning.

3 The complexities of debates around Jacobs' work are somewhat elided here, as to fully engage with them would render the example ineffective for illustrative purposes. I am aware that the narrative is not quite so simple as presented here and generally found in planning literature. For example, some of Jacobs' contemporaries, such as Herbert Gans (1968), questioned the presumption that placemaking via diversity and organic bohemianism was desirable. Indeed, Gans turned Jacobs' method of immanent critique against her by arguing that her views reflected a desire to impose an elite perspective of vibrant urban living upon the middle classes, who had a different vision of city life.

4
THE PLANNER AND THE COMMON GOOD

Moral Fiction and Reality

It was argued in the previous chapter that the work of Alasdair MacIntyre provides insights that help to clarify the underlying normative orientation of planning as a 'tradition' and 'practice' that endeavours to secure the common good of a better (or no worse) future. Here it was shown how MacIntyre identifies, discusses and questions a number of the philosophies grounding perspectives in modern ethics that have influenced planning theory and practitioner activities. The first of these to be challenged is the deontological philosophy of Kantian rationalism and forms of Lockean contractarianism that privilege rights and obligations by centralising rules within normative theory. This line of thinking has evolved from its original instantiations but weighs heavily on the philosophical justification for spatial governance where formidable concern for 'fair procedure' may sideline substantive debate on what it is that 'good planning' amounts to' (Campbell, 2006: 97). MacIntyre also criticises what he considers to be the pretensions of this perspective to ethical objectivity. In summary, he argues that this claim to independent moral truths advances ideological blindness by ignoring how seemingly objective ethical assertions are implicitly informed by the context in which they are formulated (MacIntyre, 1984: 44). He also locates pretensions to this 'view from nowhere' (Nagel, 1986) in consequentialist perspectives rooted in the utilitarianism of Bentham and Mill, whose inspiration underlies the justification for much planning activity (Campbell and Marshall, 2002). MacIntyre views as a 'moral fiction' claims that appeals to calculating the 'greatest happiness of the greatest number' can provide objective justifications for actions, especially in light of potential confusion regarding what 'happiness' entails. Hence, he is critical of declarations concerning moral objectivity, as he considers that appealing to the logic advanced by seemingly impartial deontological or consequentialist philosophical perspectives allows agents to veil their subjectively determined aims within deceptive rational language. He views such appeals as characteristic of modernity wherein moral fictions are deployed to 'close off shared deliberation about means to ends, or to close debates, or to disqualify opponents from shared deliberation' (Lutz, 2012: 99). This he sees as the 'emotivism' that has prevailed since the Enlightenment project sought a rational apologetic for the received moral traditions of a Christian worldview. To this end, his analysis of modern ethical philosophy leads him to conclude that appeals to objective methods of moral rationality constitute what Nietzsche identified as a 'will to power' in the desire to manipulate others. Thus, MacIntyre is empathetic with the plight of the secular ethicist who despondent with the veiled arbitrariness of post-Enlightenment morality opts for Nietzschean nihilism. However, while understanding this conclusion, MacIntyre is not supportive of it. Instead, he infers that such a Nietzschean view ultimately results in a moral confusion and permissivism equally as problematic as the emotivism against which it is addressed. His solution is to stretch the moral gaze beyond the philosophical justification for the rules of modern ethical conduct and reconsider the purposes those rules serve as guides for human action (Lennon, 2015b). In this way, MacIntyre finds a path out of the moral wasteland in the work of Aristotle and his concern with 'purpose' (telos). However, MacIntyre stops short of providing us with details on conceiving how one's 'moral reality' is constituted and gives shape to action. Hence, while furnishing us with tools to appreciate the tradition *of* planning as a quest for the common good, he only partially supplies the conceptual machinery necessary to understand how the situated practice *of* planning is expressed in the thoughts and actions of planners. For this, the work of the Canadian philosopher Charles Taylor is helpful.

The Right and the Good

Taylor shares many views with MacIntyre and has ostensibly referenced him as influential in developing his perspectives (Taylor, 1989). Foremost among these is the view that the modes of thinking formulated during the Enlightenment and resonant to this day have erroneously bounded the terrain of moral[1] debate. Specifically, by drawing strength from the scientific revolution, modern epistemologies rely on an 'atomistic account of putatively undifferentiated nature (including human nature)' such that they have 'erected a perniciously sharp distinction between knower and objective of knowledge' (Calhoun, 2000: 3). The result of this has been to distort and narrow the scope for understanding peoples' moral lives. For both Taylor and MacIntyre, much of this malaise is consequent on how modern moral philosophy has 'tended to focus on what it is right to do rather than on what it is good to be, on defining the content of obligation rather than on the nature of the good life' (Taylor, 1989: 3). Taylor is critical of this 'stripped-down approach' (Abbey, 2000: 11) wherein 'the focus is on the principles, or injunctions, or standards which guide action, while visions of the good are altogether neglected' (Taylor, 1995a: 145). This reliance on a 'disembodied, decontextualized, and disengaged subject' (Calhoun, 2000: 6) leads him to criticise moral theories rooted in utilitarian reasoning that have proved so influential in professional planning (Campbell and Marshall, 2002). He is likewise disparaging of the deontological deduction pioneered by Kant that is echoed in the thinking of philosophers such as Habermas (1985, 1987), whose work has informed much planning theory. Taylor calls this approach to moral thinking 'formalism' (1985a: 231) and agrees with Michael Sandel (1982) that it operates by eliding the moral suppositions upon which it is based by keeping its most basic insights concealed. As Taylor explains,

> Impelled by the strongest metaphysical, epistemological, and moral ideas of the modern age, these theories narrow our focus to the determinants of action, and then restrict our understanding of these determinants still further by defining practical reason as exclusively procedural.
> *(Taylor, 1989: 89)*

Hence for Taylor, as with MacIntyre (1984), the crux of the problem in much modern moral philosophy is a failure to acknowledge how it enjoys support not through its impartiality but by its resonance with what Rawls refers to as the 'intuitions' of the community of moral interpreters who employ it (1971: 41). As with MacIntyre, Taylor is also critical of the 'subjectivism' of those whose perspectives show a Nietzschean pedigree, such as the existentialism of John Paul Sartre (1969). In Taylor's view, the idea that a person's values matter not because they are deemed inherently worthy but rather because an individual has chosen to independently assert them is a flawed account of how people experience and act in a world of others. Hence, echoing MacIntyre, what Taylor strives for is a 'moral realism' that acknowledges a possible plurality of views on what constitutes the good but is nevertheless aware that this is a limited plurality and is not simply open to affirmation by a wholly autonomous subject. Taylor's challenge is thereby to explain how moral understandings are constrained configurations contingent on context. In this he extends the work of MacIntyre by emphasising how the contingency of one's situatedness within a shared form of life engenders a 'moral reality' wherein

> one is orientated to taken-for-granted notions of the goods of that form of life . . . that in turn structure one's practical reasoning as a kind of skill, a kind of know-how in making one's way around in that life.
> *(Pinkard, 2003: 192)*

He achieves this by grounding his standpoint on the insight of Merleau-Ponty (2002: xxii) that 'because we are in the world, we are condemned to meaning.' Specifically, Taylor maps the connections between the senses of the self and moral perspectives to show how both the procedural neutrality and sovereign subjectivism of much contemporary moral philosophy is misplaced. Exhibiting roots resonant with MacIntyre's reconception of 'tradition' and 'practice,' what Taylor engages is a philosophical anthropology that explains the phenomenology of a subject's moral life.

The 'Subject' of Moral Realism

At the heart of Taylor's moral realism is the concept of 'strong evaluation'. He takes Harry Frankfurt's (1971) distinction between first and second-order desires as his departure point in formulating this concept. First-order desires are cravings that we share with animals, such as the desire for shelter, food and freedom from danger. These desires are wholly explained by the types of behaviour necessary for their fulfilment. For Taylor, the satisfaction of such desires involves a process of 'weak evaluation'. What is important here is that the actor responds unreflectively to a want. Conversely, second-order desires are 'desires about desires, desires which enable us to arbitrate between motives and so to act in a way that is distinctive of human agency' (Smith, 2002: 89). Although Frankfurt does not use the term 'strong evaluation,' Taylor contends that the art of reflection inherent to second-order desires renders them different from first-order desires consequent on their qualitative dimension. Although we may experience a plurality of desires, strong evaluation indicates that we do not experience these equally. Instead, these desires are experienced in relationships of comparative worth. Strong evaluation thereby refers 'to distinctions of worth that individuals make regarding their desires or the objects of their desires' (Abbey, 2000: 17). It denotes the anthropological phenomenon that people rank their desires through qualitative contrast that profiles what goods to value and why these should be valued relative to other goods. In essence, it opens up space for reflexivity in responding to desires. Taylor holds this to be a central tenet of moral life. As he outlines,

> I should like to concentrate here on a particular aspect of moral language and moral thinking that gets obscured by the epistemologically motivated reduction and homogenization of the 'moral' we find in both utilitarianism and formalism. These are the qualitative distinctions we make between different actions, or feelings, or modes of life, as being in some way morally higher or lower, noble or base, admirable or contemptible. It is these languages of qualitative contrast that gets marginalized, or even expunged altogether, by the utilitarian or formalistic reductions. I want to argue, in opposition to this, that they are central to our moral thinking and ineradicable from it.
>
> *(Taylor, 1985b: 234)*

However, Taylor does not hold that every choice a person makes is the product of strong evaluation. Rather, he is proposing that some decisions entail a qualitative ranking relative to a commitment on why it is good to rank things in this way and, crucially, why it is good to believe it is good to rank things in this way. It is this reflexive dimension that is key to understanding the concept. A simple example may help clarify this distinction.

Imagine a municipal planner working on a new transport plan. A central part of this work involves weighing up alternatives regarding a new public bus route. The current draft of the plan projects the most cost-efficient route to be one that bypasses an economically disadvantaged housing estate. She is aware that this will mean that the residents of that housing estate will not now have access to public transport. Under time pressure, she ignores this issue and advances the plan for ratification. The principal criteria governing her selection appears to be expedience. The inconvenience caused by further delaying the ratification of the plan would result in a backlog of work for her to do. There seems to be little reflection evident here beyond the satisfaction of an immediate desire to manage the burden of work. Hence, this looks like an instance of 'weak evaluation'. Now suppose this same planner became concerned that the proposed bus route would curtail public transport access to low-income households and force them to use environmentally injurious private transport. Consequently, she begins advocating to her colleagues that the bus route should in fact serve the housing estate. For her, public transport represents what she understands to be 'planning for the common good' by helping to redress socio-economic inequalities through providing affordable transport for those who need it most in a manner that is less environmentally harmful than private motorised transport. In this sense, advocating for public transport reflects her understanding of the practice *of* planning as extending the very point and purpose of the tradition *of* planning. With the time spent advocating for amending the transport plan her work backs up such that she must stay late each evening for several days to catch up. In this scenario, the subject has reflected on the reasons why she should advocate for an amendment to the transport plan (social justice, environmental benevolence) and the reasons why those reasons matter (advancing the common good). Hence, this is an instance of strong evaluation in the practice tradition *of* planning.

It may be objected that the decision of the planner to advance the common good was a deontologically determined choice, and at first glance, this may seem to be true. However, as we saw in the previous chapter, what constitutes the common good *in* planning is contextually 'situated' by the spatiotemporal context in which the tradition *of* planning is positioned. In this vein, what Taylor's view indicates is that why we come to reason what is a 'good' and why we should value it is itself situated within a tradition-informed view on how it should motivate us. Moreover, Taylor's view negates a utilitarian reading by suggesting that engaging with the world from within a practice tradition renders void endeavours to eliminate distinctions of worth among desires by placing them on an equal footing (Taylor, 1985a: 17). Therefore, while one could attempt to harmonise with modern moral epistemologies by couching her choice in formulism's language of universal right action or the abstractions of utilitarian assessments, neither of these approaches adequately account for the 'moral experience' of the subject on deciding what is choice worthy and what is a worthy choice.

Accordingly, Taylor maintains that strong evaluation precludes us from experiencing all our choices equally. In contrast to the subjectivism flowing from Nietzschean philosophical perspectives, Taylor contends that in strong evaluation, the goods identified are 'not seen as good by the fact that we desire them, but rather are seen as normative for desire' (Taylor, 1985b: 120). Thus, the goods are experienced as possessing an intrinsic value. As such, our perspectives on what constitutes a good and what it is good to value are profiled by the practice tradition in which we are embedded and not simply by the intellectualised justifications that may be advanced in supporting them – albeit the choice to employ a deontological or utilitarian procedure for decision-making may itself reflect a strong evaluation regarding the desire for fairness or transparency as dimensions of advancing the common good. It is in this sense of perceiving that which is valued as intrinsically worthy of value, and of valuing it as an inherently worthy viewpoint, that Taylor identifies a 'moral realism' in our engagement with the world. As he notes with respect to strongly evaluated goods,

> they involve discriminations of right or wrong, better or worse, higher or lower, which are not rendered valid by our own desires, inclinations, or choices, but rather stand independent of these and offer standards by which they can be judged.
>
> *(Taylor, 1989: 4)*

Hence, the strong evaluation of moral realism implies a measure of value that is not dictated by mere preference but by 'an *independent* standard of worth against which the value of *de facto* desire satisfaction may be questioned' (Smith, 1997: 38 – emphasis in original). One evaluates strongly by appealing to a standard that is experienced as objectively existing beyond the simple contingencies of one's desires. In the case of planning practice, these standards are inherited, interpreted and transformed through the experience of engagement with the discipline's tradition of a quest for the common good. Such engagement helps profile what Taylor calls a 'framework'.

For Taylor, a framework or horizon[2] is the constellation of moral beliefs and commitments that gives shape to what one values and how one values it. In reflecting the two meanings of a 'frame,' these frameworks govern what is seen as an issue of moral concern while concurrently providing the interpretive structure supporting how such an issue is morally experienced. It may initially appear that modern individualism and secularity renders such a seemingly anachronistic and even quasi-religious view of morality of little benefit in understanding the contemporary moral subject. However, Taylor engages an extensive study of the history of moral concepts to reveal how such frameworks are an inherent feature in the phenomenology of moral life, be it religiously informed or otherwise (Taylor, 1989). Indeed, through this study, he demonstrates how these frameworks give meaning to one's life by supplying answers about the purpose of why one does what one does, while simultaneously affording direction as to what one should do. Therefore, 'one's framework provides guidance about moral questions in the broad sense; that is, about what it is right to do vis-à-vis others and about what it is good to be; about what is meaningful and rewarding for an individual' (Abbey, 2000: 34). Thus, to possess a moral framework is to possess a sense of qualitative distinction in which a series of basic evaluative commitments orientate one's outlook and choices. In this sense, frameworks have real-world implications through orientating us 'by offering implicit limits to choice and thereby making action possible' (Calhoun, 1991: 234). As asserted by Taylor (1989: 27),

> to know who I am is a species of knowing where I stand. My identity is defined by the commitments and identifications which provide the frame or horizon within which I can try to determine from case to case what is good, or valuable, or what ought to be done, or what I endorse or oppose. In other words, it is the horizon within which I am capable of taking a stand.

In this way, a framework facilitates moral positioning in that it configures how one constitutes and experiences the moral space one occupies. Indeed, Taylor employs the metaphor of a 'horizon' to convey how our thoughts and actions carry a sense of moving towards or falling away from the goods that give structure to our moral selves in a world of continuous questions about what ought to be done. Although he occasionally seems to imply that strong evaluation and moral frameworks are synonymous, it can be inferred from his writings that these are related rather than conflated concepts in which one's moral framework comprises an assemblage of strong evaluations. Returning to our example of the planner may help here to illustrate how frameworks influence action. Her determination to advocate for a public bus service through the low-income housing estate was consequent on strongly evaluating the morality of excluding that area from this public service. This evaluation was grounded in a moral framework comprising a constellation of beliefs regarding social justice and environmentalism, among other issues. Her framework thereby established what is an issue of concern and supplied a position relative to what she believed ought to be done. It was this orientation in the 'space of questions' (Taylor, 1989: 29) that prompted her to advocate for the amended bus route. This action represented her sense of moving towards planning's common good, a sense of movement influenced by the framework from which she sees the moral landscape of response-requiring questions unfold before her.

Aligning with MacIntyre's general perspective on the role of a tradition in affecting one's moral outlook, Taylor contends that our moral frameworks do not evolve in isolation but rather through engagement with others, particularly those we are positioned by chance or choice to learn from. Thus, central to the development of our moral frameworks is a continuous real or imagined exchange with others. He notes,

> My discovering my own [moral] identity doesn't mean that I work it out in isolation, but that I negotiate it through dialogue, partly overt, partly internal with others.
>
> *(Taylor, 1995b: 231)*

However, this complicates our suppositions regarding the planner in the above example who opts not to advocate for the public bus route. The initial assumption here is that her decision not to intervene evidences little reflection beyond the aspiration to manage the burden of work in an instance of what appears to be 'weak evaluation'. Yet, if moral frameworks are composite of strong evaluations and such frameworks develop through engagement with others, it seems logical that such moral frameworks emerge and evolve through patterns of weighing strong evaluations across different traditions that are brought to bear in different types of decision contexts. This is because the subject is simultaneously situated and stretched across multiple worlds. Accordingly, there may be multiple traditions within which a person concurrently lives and acts. These may resonate or conflict at different times in different circumstances and in different ways such that a person may be answerable to moral dilemmas that are nuanced, complex and even disempowering. Therefore, our planner who chose not to advocate for a public bus route may well be steeped in a desire to advance the tradition *of* planning in seeking to reflect upon and advance the common good of a better (or no worse) future. However, she is also a person steeped in cultural mores, familial affinities, friendship lineages and so on.[3] Each of these will have its own traditions of reflection on what constitutes doing good – being a good daughter, good sister, good mother, good friend – which may generate the need to weigh advancing one tradition of the good (e.g. the tradition *of* planning) against another (e.g. the tradition *of* parenting). Hence, while the decision not to advocate for a public bus route may appear to be made on the principle of expedience, it may actually be consequent on, for example, an overriding desire to attend to her ill child (tradition *of* parenting). Thus, as noted by Healey (2006: 47),

> typically, we live in multiple relational webs, each with their own cultures, that is, modes of thought and systems of meaning and valuing. As active agents, and in the social situations of the relations within which we live, we construct our own sense of identity. Thus we may well experience the clash of cultures within ourselves, and within the nodes of our relational webs, in the workplace, the household, the bar, the sports club, the community group.

Consequently, the planner as a person embroiled in the moral dilemmas of daily life is both a moral subject situated within the tradition *of* planning, but also someone stretched across sometimes competing traditions of the good in other dimensions of their situated existence. In this sense, the moral horizon of the planning subject is complicated by sometimes competing demands, wherein moving towards the good of one tradition means moving away from the good of another. Yet, this does not diffuse the sense that one is moving towards or away from the good of a tradition (planning, parenthood, friendship, etc.); it simply reflects a condition of moral life. Hence, how one comes to know and negotiate the moral landscape in the tradition *of* planning does not imply that one always has the opportunity to advance what one identifies as the good in one's planning activities (Tewdwr-Jones, 2002). Instead, it simply provides the horizon of moral understanding within which one gets one's bearings for weighing what *should* be done in advancing the tradition *of* planning in the broader context of what *ought* to be done relative to the other traditions that lattice one's life and what *can* be done within the political, legislative, environmental, social and economic context in which one finds oneself. Consequently, the moral landscape of planning is the horizon against which the planner interprets their position relative to the good of planning (the common good) in evaluating the degree to which they are moving towards or away from it and not as such the goodness of their life – even if the former influences an assessment of the latter. This horizon is thus one dimension in the complex moral framework informing the planner's thoughts and actions, yet the one against which the standing of their actions in a professional planning context is evaluated. Hence, what is proposed here is not an all-encompassing theory of structuration, such as that advanced by Bourdieu (1972) or the Giddensian institutionalism employed by Healey (2006). Instead, the focus here is limited to understanding the complex relationships between the planning subject, their moral life and the tradition *of* planning.

Contouring Planning's Moral Landscape

For the planner, a central moral dialogue in one's practice is with one's immediate planning colleagues and the wider community of planning practitioners (Friedmann, 1973: 171). This is because, it is 'only within a community that sustains, legitimizes, and disciplines practices and cultivates a moral order does it make sense to talk a certain way or to believe that anything is worth knowing' (Mandelbaum, 2000: 31). Hence, both MacIntyre's and Taylor's theses resonate with the 'communities of practice' theory advanced by Wenger (1998). Here, a community of practice is defined by a shared domain of interest. It is 'a system of relationships between people, activities, and the world; developing with time, and in relation to other tangential and overlapping communities of practice' (Lave and Wenger, 1991: 98). Membership of a community 'implies a commitment to the domain, and therefore a shared competence that distinguishes members from other people' (Wenger-Trayner and Wenger-Trayner, 2015: 2). Such a domain is most often conceived as a specific professional activity, such as planning (Wenger and Snyder, 2000). As a practice-based learning theory (Blackmore, 2010; Wenger, 1998), a community of practice is not simply equated with a community of interest. Rather, 'members of a community of practice are practitioners. They develop a shared repertoire of resources: experiences, stories, tools, ways of addressing recurring problems – in short a shared "practice"' (Wenger-Trayner and Wenger-Trayner, 2015: 2). Novices move into such a practice through a process of learning: first via a formal academic journey where they imbibe the community's paradigmatic truisms (e.g. public participation is good) and subsequently by working contact with existing practitioners. Through these channels, they progressively absorb the mores of that practice. Developing a sense of solidary with one's community of practice operates in a trilectic synchrony with the gradual acquisition of topic-specific knowledges and an evolving sense of professional ethics. It is in this manner that the planner emerges as a tradition-informed practitioner who knows what, how and why things are done the way they are (Davoudi, 2015). Accordingly, they come to know how to implicitly recognise the common good and plan for it, even when at times they may find it difficult to articulate what the common good is. Hence, 'if understanding makes the practice possible, it is also true that it is practice that largely carries the understanding' (Taylor, 2004: 25).

An important aspect of the solidarity, knowledges and ethical tuning acquired via interaction with one's community of practice is that this is a 'situated' experience. The insights of Taylor indicate that this situatedness has an inherently moral dimension that is shaped by its practice tradition. This is consequent on how the inherited, interpreted and occasionally transformed contours of this tradition influence one's sense of the good of planning as they help fashion the space of questions in which one seeks orientation. Guidance is found through this moral landscape in the form of ethical standards transmitted via the norms of the practice in which one engages. Accordingly, in planning, we have contextually associated conceptions of what constitutes planning for the common good and as a corollary what entails proper practice. For example, in a politically authoritarian state, prevailing understandings of planning for the common good may differ from prevalent understandings in a more democratic context. The former may commonly witness the deployment of command-and-control approaches to identifying issues and actioning responses, while in the case of the latter, seeking greater public participation in decision-making may be the norm. While utilitarianism and formalism may be employed in different ways in both contexts to ostensibly underpin the series of decision-making processes evident and the decisions made, it is the tradition-informed practice *of* planning that profiles the ethical assumptions supporting views on 'what' ought to be done and 'how' it ought to be done. Such profiling thereby facilitates the identification of issues of moral concern and supplies the impetus for action. For this reason, we can have the scenario apparent in various contemporary contexts where opinions on what constitutes planning for the common good can differ considerably. For example, in certain contexts, it can be deemed morally agreeable from a practitioner standpoint and ethically acceptable from a practice perspective to appropriate private property from poor residents for the construction of a privately owned retail development that will serve the consumption aspirations of a burgeoning middle-class. In such a tradition-informed practice context, this regressive redistribution of an asset can be perceived as morally tolerable by a planner consequent on a qualitative distinction between the needs of some members of a population relative to the desires of others (they're only poor, live-in shacks, are dragging this area down and can easily move on; conversely, the middle-class bring money, civility and will help improve this area). Furthermore, this may be ethically legitimated via a utilitarian argument that equates what ought to be done with the greatest benefit accruing to the greatest number consequent on the economic activity ensuing from the construction of the new retail development. In contrast, in a more democratic context such flagrant disregard for resident views simply on the basis of their socio-economic standing would be considered morally odious and ethically unconscionable: a view that is contrary to understandings on how the tradition *of* planning for the common good should be realised. In the context of modern western democracies, such a view is liable to be expressed through deontological practice ethics. However, what is of note here is that neither perspective represents an 'objective' identification of planning for the 'common good' as each perspective is contextually contingent on the tradition-informed practice that directs action on how to weigh up a complexity of various potential impacts against the backdrop of a schema of ethical assessment on what ought to be done. While a view from nowhere is therefore impossible resultant from the fundamental paradox of social life that 'there is no way to fix neutrality neutrally' (Margolis, 1998: 59), this does not erase the sense held by the planner and their community of practitioners that the views they hold on right and wrong are real and independent of their localised application. Indeed, it is this sense of moral realism that underlies what Taylor refers to as the 'social imaginaries' that organise our world (Taylor, 1989).

As a community of practitioners concerned with the tradition *of* planning, planners engage with a world of contending agendas in a way that must make sense of complexity to effectively make decisions. Taylor coins the term 'social imaginary' to convey the tacit patterns of sense-making that furnish the 'common understanding' that renders possible a shared practice by providing it with a self-aware sense of moral legitimacy (Taylor, 2002b). Such a social imaginary encompasses the implicit norms of social existence and supplies the set of expectations of the world that make such norms realisable (Taylor, 2002a). It is more than just a suite of ideas in planning. Rather, it is both the basic shared hermeneutic through which we interpret normative relations in the world and simultaneously a prescriptive force limiting the plurality of views on right and wrong. In this sense, a social imaginary is a meta-structure that profiles and is profiled by frameworks in much the same way as frameworks shape and are shaped by strong evaluations. Accordingly, it delineates the very range of moral possibilities in which planning as a normative endeavour can legitimately operate. As described by Taylor,

> by social imaginary, I mean something much broader and deeper than the intellectual schemes people may entertain when they think about social reality in a disengaged mode. I am thinking, rather, of the ways people imagine their social existence, how they fit together with others, and how things go on between them and their fellows, the expectations that are normally met, and the deeper normative notions and images that underlie these expectations.
>
> *(2004: 23)*

As such, a social imaginary encompasses a sense of factuality where the norms of interaction and expectation exist in a moral realism that enables the solidity of a shared practice in a complex and contingent world. Taylor asserts that 'implicit in this understanding of the norms is the ability to recognize ideal cases . . . . And beyond the ideal stands some notion of a moral or metaphysical order, in the context of which the norms and ideals make sense' (Taylor, 2004: 24–25). Thus, in the case of our example of an authoritarian state facilitating a private retail development by evicting poor residents from their land, the social imaginary of modern western democracies that turns on the concept of equal and inalienable human rights as inherent to advancing the common good of planning would lead the planning practitioner to interpret this as morally wrong and ethically unacceptable. To condone such action would be to move away from the common good by diluting the principles of human dignity. Hence, for the planner schooled in the contemporary western practice tradition *of* planning, the common good is likely to involve promoting public participation as a means to facilitate dialogue in collaboratively devising a solution acceptable to all (Healey, 2012b; Innes and Booher, 2003) or to mediate between conflicting viewpoints in an effort to reach agreement on a way forward (Forester, 2009).[4] Conversely, the planner from the authoritarian state may be operating within a social imaginary wherein a legacy of autocracy and strong communitarianism dominates. For her, planning for the common good justifies responding to the perceived greater needs of the wider municipality by discounting the needs of a minority. Such 'greater needs' are determined through the expertise of the professional planner schooled in the technicalities of rational decision-making on complicated issues. To seek the involvement of an uneducated minority in the professional process of adjudicating on the right course of action would be contrary to the practice tradition *of* planning in advancing the common good, as it would wrongly privilege the voice of an ignorant minority at the expense of the many. In essence, what differs here are the principles underlying 'the ways people imagine their social existence, how they fit together with others . . . and the deeper normative notions' that sustain expectations (Taylor, 2004: 23). By emphasising different ideals that legitimise different norms, the principles that fashion different social imaginaries influence the frameworks that they mould. This then helps shape the evolutionary path that the tradition *of* planning takes. Therefore,

> in conceptualizing the idea of planning, the planning subject has to be related to and questioned in a social context. Planning is not a matter of free invention or construction but *reconstruction* with its re-invention conditioned and enabled by existing norms, positions and social expectation.
>
> *(Salet, 2018a: 24 – emphasis in original)*

Notes

1 Taylor uses the terms 'moral' and 'ethical' interchangeably. As will become clear, I use these terms in their more conventional academic form: 'moral' indicates personal views along a spectrum of what is right and wrong; 'ethics' indicates those views held by a community of interpreters. The latter thereby refers to professional conduct, while the former indicates one's beliefs on qualitative distinctions.
2 Taylor uses these terms interchangeably.
3 I would like to thank an anonymous reviewer of an earlier draft of this chapter for drawing my attention to this issue.

4 Of course, some theorists would argue that such introductions may deflate calls for meaningful public involvement in decision-making through 'a system focused on carefully stage-managed processes with subtly but clearly defined parameters of what is open for debate' (Allmendinger and Haughton, 2012: 90), which thereby may ultimately reinforce exclusionary forces rather than addressing them.

5
ADVANCING THE COMMON GOOD

Tracing Change

The conceptualisation presented thus far is quite abstract. Hence, this chapter seeks to illustrate how the tradition *of* planning and the moral framework guiding a practitioner's outlook can interact in ways that reorientate a practice *in* planning in a manner conditioned by, but transformative of, existing perspectives on what best serves the common good. The illustrative case study drawn on is the formulation, dissemination and integration of a new planning approach to reconciling nature conservation with human activities in planning for green space multifunctionality. This may seem an odd choice given that most current debates on the common good in planning centre on 'social' concerns, such as equity, diversity and enhancing public participation in decision-making. To some extent, this can be attributed to a contemporary emphasis on issues of 'justice' in the tradition *of* planning. Nevertheless, how to address social-ecological interactions in a world of expanding urban areas and depleting ecosystems integrity is increasingly extending the moral gaze of planning beyond what we *can* do to what we *should* do in creating a better (or no worse) future. Indeed, the very nature of nature conservation means that as with planning, it is an inherently normative tradition – a fact often obscured in the scientific language of ecology and the legalese of regulation. This is because at its core nature conservation is 'a collection of social values and agendas that seek both to protect species and natural areas, and more generally, to govern our relationship with the natural world' (Jepson and Ladle, 2010: 17). In this sense, nature conservation responds to how we think we 'should' orientate ourselves in a world of moral questions on what 'should' be done and why. It thereby seems reasonable that examining the transformation of land-use planning regarding nature conservation offers insight into how a moral framework orientated to the common good can prompt an evolution of the tradition *of* planning by reorientating practices

DOI: 10.4324/9781003155515-8

in planning. Accordingly, this chapter investigates how the approach to planning for nature conservation and green space was transformed via dissemination and integration of a new planning concept called 'green infrastructure' (GI). Specifically, the story narrated follows the emergence and evolution of the GI concept in Ireland between the years 2008 and 2011. This provides a fitting case to illustrate the intersection of moral frameworks with the tradition *of* planning consequent on the relatively short period in which the concept emerged from obscurity to widespread statutory representation in policy documentation at multiple levels of planning governance. Furthermore, Ireland's relatively small population of just 4.6 million at the time of the concept's emergence (CSO, 2011) is reflected in the limited number of actors with the power to institutionalise the GI approach into statutory land-use governance. This comparatively narrow temporal and administrative frame thereby renders it feasible to comprehensively chart the path taken in the rise of GI and its placement on the policy agenda.[1] It also makes it possible to confidently identify the roles played by different agents in using the concept to extend the tradition of planning as that which seeks to advance the common good.

So what then is this 'GI' concept? As noted by Mell, 'Green infrastructure planning has come a long way in a relatively short period of time' (Mell, 2019: 9). While the origin of the term remains debatable (Lennon, 2018), and there are a variety of interpretations as to what it entails (Sinnett et al., 2015), nearly all understandings resonate with the frequently referenced definition advanced by Benedict and McMahon (2006: 1) as 'an interconnected network of natural areas and other open spaces that conserves natural ecosystem values and functions . . . and provides a wide array of benefits for people and wildlife.' The GI concept has been adopted across a broad spectrum of countries, from those in North America (Austin, 2014; Richards, 2018; Rouse and Bunster-Ossa, 2013) and Europe (Kabisch et al., 2017; Pearlmutter et al., 2017), to India and China (Mell, 2016, 2019). Given this array of geographic locations and the variety of planning systems in which it is used, it is unsurprising that GI is a continually developing concept whose meaning is often dependent on who is employing it and the context in which it is deployed (Lennon, 2019). The emergence of the GI concept in Ireland is no different (Lennon et al., 2016). Although largely unknown among the Irish planning fraternity prior to November 2008, the GI concept enjoyed a meteoric rise in popularity among spatial planners and allied practitioners subsequent to that date. This was consequent on a desire by some to change long-standing practices at the intersection of planning and nature conservation that seemed to ignore the socio-ecological relationships that connect people to their environment. For planners and others interested in change, conventional planning practice served to partition areas for nature conservation from areas of recreational amenity and flooding management, with little attention allocated to nature in the built environment. Such practitioners came to believe this was contrary to the common good by undermining the long-term objectives of nature conservation. In addition, many of those seeking change variously felt that current modes of nature conservation planning were diminishing the quality of the public realm, generating unnecessary costs for the exchequer and reducing the health and well-being opportunities afforded to people by contact with a biodiverse environment. Hence, what prompted a desire for change was concern regarding the configuration of values, agendas and relationships with respect to how the issue of nature conservation was been handled in planning practice. In this sense, the tradition *of* planning was advanced through the practice *of* planning that resulted in the transformation of practices *in* planning.

Transforming Practices *in* Planning

Although GI was first mentioned in an Irish context by Tubridy and O Riain (2002) as a scientifically focused 'ecological networking' concept, most planners in Ireland attribute Fingal County Council (FCC) as the initiating source and one of the principal advocates behind GI's ascension to prominence in the Irish planning system. Furthermore, it is widely held that FCC's 'Heritage Officer,'[2] Dr Clabby,[3] was key in championing the idea of GI both within the council and in the Irish planning system more generally (Lennon, 2015b). In relaying his reasons for advocating GI planning, Dr Clabby conveys a narrative of reflective transformation wherein his efforts to promote a GI approach convey what MacIntyre and Taylor refer to as an unfolding 'quest' to advance the tradition *of* planning for the common good. To Dr Clabby, this story begins with his training as an ecologist and his subsequent university lecturing career. As he notes,

> I suppose when you come in later on into somewhere like a Council [planning authority], you bring whatever you have with you, and you are more likely to articulate it clearly. Like part of the reason I wanted to get out of academia was that I wanted to be involved with somewhere where I was 'doing', trying to influence conservation or that agenda in reality, rather than lecturing other people about how you might do it.

Through this statement, Dr Clabby furnishes a concise explanation of how his decision to move from academia to practice was propelled by a desire to move towards the good of nature conservation by influencing how it is realised in practice. As 'a collection of social values and agendas' (Jepson and Ladle, 2010: 17), nature conservation represents for Dr Clabby something that is choice worthy. This is set within a moral framework that conceives a career path orientated towards the goal of influencing the conservation agenda as a worthy choice. Facilitating his desire to 'influence conservation' was the flexibility to interpret and advocate the broadly conceived idea of 'heritage' in the setting of a local planning authority. As he reflects,

> Each of us [Heritage Officers] have created our own role in some ways, because there was no template for this job, like when I came here there wasn't a Heritage Officer, so you just create your own role I suppose, which gives you great freedom in some ways.

While this 'freedom' provides flexibility in advocating for the greater allocation of attention to heritage matters, what Dr Clabby discovered on entering FCC was an operational context that he regarded as inadequate with respect to nature conservation. As he recalls,

> One of the shocking things for me when I came from a university into a local [planning] authority, and it really shocked me, was that people ignored the law, even in the local authority. I couldn't believe it. I thought 'it's a SAC,[4] its legally protected. Like what; you're just ignore[ing] it?' That was happening all over the place! And that was kind of a shocker. But then you realise this is Ireland; we pass all these laws with very little intention of wondering whether anyone keeps them or not. And so, then you could see the law wasn't going to work; a few people rabbiting on about these issues wasn't going to work.

Hence, Dr Clabby found himself 'situated' in a context that ignored appeals to conventional legal instruments used in nature conservation. He thereby came to realise that the typical (legal) tools wielded by those seeking to advance nature conservation were likely to have little impact in a local planning authority setting. Confounding this widespread flouting of environmental law was a general lack of attention shown to nature in planning policy formulation. As he notes,

> We figured out where to put residential stuff, where to put commercial things, where to put roads, where to put everything and then if there's any green stuff left over, well sure we'll have a bit of that for a park. And I just don't think – that's not good enough!

This failure to adequately consider 'green stuff' (nature) is cogently identified as a deficiency in planning. This defective approach distinguished through a moral framework from where a qualitative distinction on the value of nature relative to how contemporary practices *in* planning accounted for it positioned Dr Clabby in a 'space of questions' (Taylor, 1989) of 'what' was wrong and 'why' it was wrong. For Dr Clabby, this would thereby necessitate finding new ways of thinking and doing should he seek to advance nature conservation within a local planning authority context. In essence, it would require locating a means to transform practices *in* planning that failed to allocate appropriate attention to nature. Accordingly, as suggested by the above set of quotes, there occurred an implicit movement towards the good of nature conservation from 'trying to influence conservation' to knowing what won't work. Set against this realisation, Dr Clabby sought alternative means to promote nature conservation in planning practice. It is therefore somewhat ironic that when reflecting on his first contact with GI, Dr Clabby noted that this encounter, although self-motivated, was initially unexceptional,

> How I came across this concept was when I came here [FCC] first in 2003 they were doing the development plan at the time, which was adopted in

2005, and [I] just arrived after the first stage of that and the first job was 'write the natural heritage chapter for the development plan please', and I was Googling things and I found this green infrastructure paper by Benedict and McMahon[5] on the internet and I read it and I thought that's interesting, don't have time to think about it now but I'll file it away in my head and I'll think about it later.

However, it appears that as time progressed, Dr Clabby became aware of what he considers the benefits of the GI approach outlined by Benedict and McMahon (2002). He explains the development of such ideas when conveying his observations on what he perceives as the low profile credited to biodiversity issues in planning that conceived nature conservation as a problematic issue to be negotiated in the context of other, more important issues, rather than an important issue that should be advanced in itself. As he recalls,

So you know SAC's were very much viewed at the time as being something there that somebody else designated . . . but the problem with that approach was that then they [planners] were zoning land right next to them and then having an issue when somebody like me came along and said well 'now there's an issue', when you want to do something in the zoned land. So the Benedict and McMahon kind of formula of saying, well look it's about thinking early and it's about integrating these things and seeing nature conservation or other functions, like as kind of real things that you need to provide for in a planning context, you can map it, you can call it 'infrastructure' – makes it seem important. It doesn't say things like 'biodiversity' which people don't seem to understand.

Here, Dr Clabby expresses the view that he came to see GI as a means to address the perceived low status of biodiversity issues in planning policy formulation by offering a profile-raising communicative means to address the poor integration of nature conservation into planning activities. This was undertaken as a response to the perceived failure of the more scientific and unfamiliar term of 'biodiversity' to convey the importance of nature conservation to a non-scientist audience. Hence, GI was both initially perceived and employed as a means by which to articulate the importance of nature conservation, and subsequently elevate its status in planning policy formulation (Lennon, 2015a). Achieving this was enabled by the perceived usefulness of GI in resonating with forms of justification familiar to engineers and planners as to what constitutes a common good ('infrastructure') and the connotations which flow from this on how that common good can be delivered through the professionally familiar activities of planning and designing 'development' (Lennon, 2015c). As he reflects,

The thing that attracted me about it [GI] was that it made sense. It was a language that seemed to me to make sense to the likes of engineers and planners . . . to me nature conservation people were just not talking a language that other people understood. . . . [The] notion of development and conservation going hand-in-hand seemed to me to be absolutely the way to go. Because constantly what I could see all the time I was being confronted with, in a development management context [assessment of development proposal applications] . . . were these conflicts all the time because people hadn't been thinking early enough about the issues.

According to Dr Clabby, his appreciation and interest in GI as a concept was intensified following a further investigation into GI as a planning concept several years later. In particular, he became aware that GI was a discourse existing beyond academic speculation and enjoying some popularity among planning practitioners in foreign jurisdictions, notably in the UK and North America. This appreciation was subsequently consolidated during a study tour of the Dutch 'ecological network'. In this respect, Dr Clabby notes,

I was very impressed by their [the Dutch] kind of thinking which was that, 'look for years nature conservation in the Netherlands, we just talked to ourselves and we never got anywhere and we figured out we needed to talk to people who didn't want to talk to us but we needed to find a way of talking to them', like the farming people, like the water people who manage their waterways, their developers. So they'd done that by developing this whole thing about ecological structure and knowing what they wanted and presenting it and getting it through. Yeah I thought, that's really cool. I got excited about it and thought, yeah we can do that, why don't we do it.

Dr Clabby's emerging perception of a necessity to accommodate human land use needs within nature conservation planning to both facilitate biodiversity protection and the planning of more agreeable environments for people appears to have been heavily influenced by the coincidence of this study tour with an increased realisation that GI was a planning approach adopted in several jurisdictions outside Ireland. Moreover, how it was rationalised in such jurisdictions furnished Dr Clabby with a solution to what he perceived as the problem of nature's neglect by Irish planning at that time. This seems to have instilled a desire to introduce GI planning in Ireland as a means to address issues surrounding human interactions with areas of ecological sensitivity. As he notes,

We have to get beyond regulation, you know, I mean I would be of the view that there's no point in things like the Natura 2000 network of sites, in just seeing it as a regulatory job, like that's a road to nowhere. You know we put way too much emphasis here on the regulation of these things at the moment and not enough emphasis on their potential and their ability to kind of build-up community and to be places that people enjoy.

We see here a movement from Dr Clabby's initial 'shock' that nature conservation regulations were not being adhered to by local planning authorities to a view that such regulatory instruments were ultimately of little benefit in advancing nature conservation, as their application tended to overlook the capacity of nature conservation activities to enhance social and place-based affect. In essence, his appraisal constitutes a critique of the appropriateness of the approaches taken to nature conservation in contemporary planning. This developing perception of such approaches as 'a road to nowhere' and the discerned need 'to get beyond regulation' points to the contextual 'situatedness' of Dr Clabby's evaluation of the modes of environmental planning operative at that time. It also indicates an ethical assessment informed by a moral framework on the need for change. This thereby propelled him to seek a transformation of the practices *in* planning regarding nature conservation. This view was intensified by Dr Clabby's view that, 'ecologists generally who all understand why biodiversity conservation is important . . . have not been very good at communicating this to a wider audience.' This assessment was lucidly conveyed when stating,

> I felt well what's the point in us [ecologists] rabbiting on about this stuff and going to conferences where the only people we're talking to is ourselves . . . and when the planners pull down a map and know what they want, we never know what we want other than 'protect that thing there.'

Thus, by operating within and between the communities of practice of ecologists and planners, Dr Clabby was able to identify the deficiencies in both on how the issue of nature conservation was approached. For him, key to advancing the common good of nature conservation was transforming the ways in which nature conservation was handled in the planning system. This involved a shift in modes of justification from that solely privileging scientific expertise to new ways of thinking and communicating the value of ecosystems. Such a shift entailed a move from the existing practice *in* planning of predominantly reactive nature conservation consequent on the scientific assessment of the impacts of a development proposal. This involved the reframing of nature conservation away from an overreliance on reactive instruments towards a new modus operandi focused on maintaining and enhancing ecosystem functions rather than just minimising damage. As noted by Dr Clabby when recalling issues concerning the reliance on environmental impact assessment (EIA) as the principal means to address nature conservation in planning,

> Even that whole premise if you like of EIA's, which is, maybe the premise is that the environment is ok and that . . . all we have to do; our job is just to make sure we don't damage it. And I kind of feel that we're gone way beyond that now. Like human beings are in control basically. It's not like the environment is out there doing its own thing; it's not. So, it's not about 'are we doing any damage?' It's about 'how can we do what we do and enhance or protect . . . the environment?' . . . . It's not just about 'protecting'. Like I think that's a

very 60s kind of idea in some ways, which is where the EIA came out of, the 60s: is that the environment is OK and we must not just damage it. Like I think there's 6 billion of us or whatever number of us now on the planet. Nature conservation is totally changed. It's now about ecosystems services. We've moved into a different place. It's not just about making sure everything is OK and we're not damaging it. So I think that 'paradigm' if you like, so the 'green infrastructure paradigm' if you like, or 'ecosystems services paradigm', is about how do we continue to provide these viable functions for society while doing what we need to do, like building a road? Not just how do we do minimal damage, which seems to me a very kind of not satisfactory way of framing things.

GI as a new way of 'framing things' would thereby mean revising the way issues, things and people are identified as significant and considered in the formulation of policy and the determination of development proposal applications through a new way of thinking about how the common good of nature conservation is best realised. As such, it would necessitate a transformation of contemporary practices *in* planning regarding nature conservation. This would entail a reprofiling of those justifications used in formulating nature conservation. In essence, it involved moving beyond ecological expertise to incorporate the ways other communities of practice think about value in seeking to realise the common good. Through his experienced-informed critique of the deficiencies of ecological and planning approaches to nature conservation, Dr Clabby thereby became convinced of the need to broaden the community of those informing policy development and decision-making regarding nature conservation in a local authority setting. Thus, rather than remaining heedless to the land-use aspirations of non-ecologically focused professions, he increasingly thought it necessary to foment support among a coalition of actors across various communities of practice that shared an interest in planning for green space and nature conservation. As he states,

> I can get on board landscape architects and parks people and maybe people who have a walking and cycling agenda, and so it's not just about me on my own arguing my little corner, but it's about making that argument stronger by finding fellow travellers who think this kind of language if you like. So I felt the green infrastructure thing had that, all of that going for it in the sense that it looked positive . . . it was a language planners and spatially oriented people could understand.

Therefore, by addressing issues across various communities of practice, GI advocacy was interpreted as a means to remedy an array of planning policy issues and thereby augment support for its use as a means to transform current practices *in* planning regarding green space provision, mobility and nature conservation (Lennon, 2015a). It was against this backdrop that GI re-emerged as a planning discourse in

Ireland at the GI conference in Malahide in November 2008. As recounted by Dr Clabby,

> We had a thing in the heritage plan saying we had to have a major conference every few years. So I thought, well let's have a conference. And for a while I was going to call it green spaces . . . so eventually I kind of took the courage of my convictions and said no, we'll call it 'green infrastructure', even though no one knew what it meant.

This conference represented GI's re-introduction to planning policy debates in Ireland from initial (largely failed) mooting as a potential scientifically grounded approach to conservation (Tubridy and O Riain, 2002). Thus, Dr Clabby's role in this re-introduction process was central. Moreover, this conference consolidated a growing interest in the GI approach among management and planning staff in FCC. Through using the conference as a platform to expound the benefits of a GI approach over contemporary practices *in* planning, Dr Clabby was able to illustrate to an assembled audience of planners and allied professionals that GI was a means to address issues they were experiencing in their current planning activities. In this context, Dr Clabby recalls how the conference was a major turning point in getting 'buy-in' from colleagues for the employment of the approach in planning policy formulation. As he recalls,

> So for me locally what it did was, people turned around like after the conference and said well 'let's do this in the new development plan'. Like just like that; bang! I didn't have to do anything else. Like that was the end of that discussion. You know because I had been kind of saying, 'we should do something about green infrastructure in the new development plan', but after the conference people just went 'yeah right absolutely'. There was no argument about it. 'Like let's go and do it'. That was great locally.

Indeed, a direct result of the conference was a desire to transform FCC's traditional mode of policy formulation for the development plan that was being drafted at the time. This involved a radical departure from the conventional format of development plans in Ireland at the time, which normally confined nature conservation and green space issues to respective 'natural heritage' and 'green space' chapters. FCC replaced this format with a structure that placed GI front and centre as a concept structuring the overall objectives of the plan, coordinating the policies of other chapters and threading through policy on land use, from issues as varied as archaeological heritage management and community development to drainage management and transportation. As Dr Clabby recollects,

> One of the critiques that came back from the planners who were using the previous plan, was that . . . the chapters were all seen to be too stand-alone, and no one was making links, or it was hard for a planner to make a link. I

suppose we've worked hard at this plan to keep referring people to other bits where we feel that that's an important thing. So, it's to try and say well look if you're thinking about design and the urban environment you have to think about green infrastructure. If you're thinking about biodiversity you need to think about green infrastructure. We're trying to make those links for planners who are using it. Because I think in some ways, development plans have kind of become very unwieldy. Like there's millions of policies and objectives under buckets of headings. How on a day-to-day basis do you actually use that document? So I suppose the fact that that came up as feedback on the previous plan helped the idea that you would even have a chapter coordinating a lot of stuff under the heading of 'green infrastructure'. People did see this kind of idea of guiding people around the place and making connections for them as good.

In this way, the GI concept became a means of ordering planning issues and those of related disciplines in the practice of weighing up what mattered where, when and why. Furthermore, outside of FCC, the conference was key in transforming planning approaches to green space and nature conservation in Ireland through the integration of the GI approach into the Regional Planning Guidelines for the Greater Dublin Area. This was important as planning authorities must ensure that their development objectives are consistent, as far as practicable, with national and regional strategies.[6] Effectively this means that once adopted as an approach within Regional Planning Guidelines (RPGs), all local planning authorities within that region were also obliged to adopt the GI approach. As noted by Dr Clabby,

> One of the real gains out of it [conference] was that [planner's name] who I had known from planning here, but then was writing the RPG's – so was the planner in charge of the RPG's – came and then suddenly they were writing a chapter on green infrastructure in the RPG's. And that's really important because now other counties in the region have to go and do it.

Additionally, the conference was important in stimulating interest in the GI approach beyond the planning profession and among allied professions. This was consequent on Dr Clabby's advocacy of the GI approach to the chairperson of the Urban Forum on the wings of the conference. From a strategic perspective, the Urban Forum represented a useful channel to disseminate the concept as it is a joint initiative by the five institutes representing the built environment professions in Ireland – namely, the Royal Institute of Architects in Ireland, the Society of Chartered Surveyors, Engineers Ireland, the Irish Planning Institute and the Irish Landscape Institute. As recollected by Dr Clabby,

> And then the Urban Institute [Forum] people. You know I'd said to [Chairperson of Urban Forum], 'look let's not let this thing die after this conference', and he said, 'oh well yeah, the Urban Institute [Forum] are a bit interested'. So

later then they had one of their meetings about it, so . . . yeah it put it on the agenda somehow. Other people began to think about it and the terminology popped up in other places.

Indeed, the Urban Forum working in conjunction with the Institute of Ecology and Environmental Management subsequently published a document for its members entitled 'Green Infrastructure: A Quality of Life Issue' (UF and IEEM, 2010) that sought to advocate the benefits of a GI approach for planning and design. Similarly, it was this conference that stimulated Comhar (the Irish Sustainable Development Council) to become interested in the GI approach. It would later seek to advance the GI approach in planning through the production of a large report on how to deploy the GI concept using cartographic methods (Comhar, 2010). Indeed, following the conference, Comhar commissioned a commentary piece by Dr Clabby on GI as a means to enhance the profile of the approach among the broader environmental policy community. As Dr Clabby recalls,

> I had asked [Chairperson of Comhar] to chair one of the [conference] sessions. He seemed to be quite taken by the thing [GI] and then I wrote this thing for Comhar afterwards. He asked me would I write a short thing for Comhar and they got a bit interested in it then.

As noted by Dr Clabby, this trend of advocating the GI approach to planning continued,

> I kept at it. I wrote things or I talked to people or if people ask me to give a talk I did it. And I suppose because I had a background in lecturing, you know I was happy to do all that stuff.

Following rapidly growing interest in the concept among the planning and allied professional community, Dr Clabby was asked by the then minister for planning in 2011 to meet with him and his aides to discuss the GI concept and how it could be used to help inform national guidance on local area planning and design that were being drafted at the time. Dr Clabby obliged and used the opportunity to explain his views on how the GI concept could lead to better practices *in* planning. When the new guidelines were formerly published, they specifically advocated the benefits of GI, noting,

> The adoption of a green infrastructure approach can contribute greatly to the effectiveness of a local area plan. This approach was introduced to Ireland by means of an international conference organised by Fingal County Council in November 2008. In August 2010, Comhar, the Sustainable Development Council, produced a research report focusing on Green Infrastructure, entitled 'Creating Green Infrastructure for Ireland'. This report sets out a broad

definition of green infrastructure and proposes an approach and a set of principles that should be followed in green infrastructure planning.

Green infrastructure has already been integrated into several strategic planning exercises including the Regional Planning Guidelines for the Greater Dublin Area 2010–2022, Fingal Development Plan 2011–2017 and Dublin City Development Plan 2011–2017. The integration and incorporation of a Green Infrastructure approach, including an initial inventory of green resources, into the local area plan process can contribute greatly to the quality of the environment in the area covered by the local area plan, to the conservation and enhancement of green resources over a wider area and to climate change mitigation and adaptation.

(DoECLG, 2013: 35)

Hence, Dr Clabby's position in the advocacy of the GI approach in Ireland was key. From an initial desire to 'influence conservation,' Dr Clabby leveraged opportunities presented by his position to rethink how planning approached nature conservation. Such work ultimately resulted in the transformation of planning at the local, regional and national levels by forging solidarity within and between built environment and ecological communities of practice in a way that prompted the generation of new forms of knowledge and judgements on how the common good is best advanced. Propelling this was an experience-informed reflective perspective grounded in a moral framework that reconceived how nature conservation could be better realised. As he muses,

I'm not a total optimist about everything, but you have to be optimistic. I mean there's no point being in this business unless you're optimistic and that things can be better in the future.

Extending the Tradition *of* Planning

As illustrated by this case study, the work undertaken by Dr Clabby to introduce GI to Irish planning was propelled by a process of strong evaluation on what is choice worthy and what is a worthy choice. This was orientated relative to his moral framework in which nature conservation was highly valued. Here, the transformation that emerged from his advocacy did so from a critique of the deficiencies of planning and the appropriateness of the perspectives circulating among ecologists, planners and allied professionals. For Dr Clabby, the GI concept offered a way to address such shortcomings, and as such was perceived as a means to move closer to the common good of nature conservation that he believed planning ought to be advancing. As a subject situated within the communities of practice of both nature conservation and planning, Dr Clabby's decision to advance the GI concept thereby emanated from a position of substantive reason from within these traditions. Hence, his decision to advocate for a GI approach to planning was born of an intersubjectively constituted 'moral subject' engaged in the practice *of* planning regarding what best advances the common good in a particular context. In this sense, his activities were propelled neither by a deontologically or utilitarian moral register. Rather, they were motivated by a moral horizon wherein moving planning towards a position of greater practice excellence in representing the common good involved moving towards a sense of the good as specified by his moral framework. This is thereby a story of transformation consequent on an immanent critique. Here, the modus operandi was challenged and ultimately changed by questioning what was being done and how this could be improved to advance the point and purpose of planning as a force for the common good. While Ireland's GI story could be read as simply a policy approach 'whose time has come,' this would be to attribute agency and inflate the inevitability of a concept while deflating the reflective capacities of those who advanced it. In essence, it would be to miss the point that GI was promoted as a means to advance the quest for better planning as a morally contoured sense of what is better/worse, which is born of moral frameworks wherein response-requiring questions emerge across a horizon of values on what it is right to do and why. Hence, the commonness of the common good of nature conservation in planning is not in question, as this is a 'fact' now held by almost all planners schooled in the universities of modern western democracies. In this sense, it is an instance of moral realism that forms part of the backdrop to practices *in* planning. It is thus 'how' this common good is realised in planning practice as an extension of the tradition *of* planning that this case speaks to. Therefore, Ireland's GI story is one where reflection on the practice *of* planning helped extend the tradition *of* planning by transforming practices *in* planning. It is a story of transformation that portrays how the effort to effect change was motivated by a moral framework situated in a particular context dominated by inherited perspectives on nature conservation, planning policy formulation and processes of decision-making.

Notes

1 This chapter draws upon an analysis of semi-structured interviews with 52 participants from the Irish public, private and voluntary sectors. This information was supplemented by the scrutiny of information obtained from participant observation at two GI-related planning workshops in Ireland and the detailed examination of 131 Irish land-use policy documents.
2 The 'heritage officer' position was developed by Ireland's Heritage Council from a small pilot programme commencing in 1999 and subsequently expanded to include officers in almost every local planning authority. Underpinned by government commitments in the National Heritage Plan (DoAHGI, 2002) and supported via shared funding arrangements between the Heritage Council and participating local planning authorities, the heritage officer programme aimed to ensure the presence of heritage expertise within the local governance system. Working from a broad definition of 'heritage,' these officers help coordinate and provide input into numerous council activities ranging from issues concerning nature conservation through to geological heritage and archaeology, as well as built and cultural heritage matters. As such, their activities interact so frequently with the local planning process that they have often become central in the processing of development proposal applications and the formulation of new planning policy.

3 Dr Clabby was the FCC Heritage Officer for 15 years beginning in 2003. He left FCC in 2018 to take up a senior position with the National Parks and Wildlife Service.
4 'SAC' is the acronym for 'Special Area of Conservation,' which is a nature conservation designation under the provisions of the European Union Habitats Directive, 1992.
5 Benedict M and McMahon E. (2002) Green Infrastructure: Smart Conservation for the 21st Century. *Renewable Resources Journal* 20: 12–17.
6 Section 7 of Part 2 of Statutory Instrument No. 30 of 2010: Amendment of Section 10 of the Principal Act.

6
PLANNING FOR THE COMMON GOOD

Seeing From Within

As an activity operating in a professional world where authority is normally predicated on the projection of detached expertise, much of the history of planning theory and practice has been profiled by attempts to ground its sense of legitimacy on explicitly impartial reasoning on how to advance the common good or its contemporary synonyms. While debates in planning theory have challenged this approach, planning practice remains significantly influenced by it. Indeed, the persistent assumption that correct procedure will deliver the common good in practitioner activities reflects a moral epistemology inherited from attempts to partition the inherent subjectivity of planners from the machinery of decision-making. However, this view fetishises an overly rationalistic account of how we reason the common good by implicitly proposing that getting the common good right is by definition achieved through following the right process and/or referencing universal moral truths. What this elides is that the nature of planning means that the very choice to select any form of adjudication is reasoned from within a communally contoured sense of ethical appropriateness rather than being removed from it. In this way, such a decision is 'intrinsically linked to the content of its deliverances and only indirectly to the procedure that generates them' (Smith, 2002: 106). Thus, the order of priority between the moral decision on what is the common good and how it is identified actually operates in reverse to that presumed. In this sense, the decision to undertake a collaborative planning exercise or advocate a power-sensitive critical-theoretical approach to research represents an instance of 'strong evaluation' on how planning should be done in the common good, which in itself is informed by the moral frameworks influencing the constitution of ethical perspectives propounded by universities,

DOI: 10.4324/9781003155515-9

planning authorities and professional institutes that carry the constellation of reasons profiling perspectives *in* planning theory and practices *in* planning. Thus, theorists, planning practitioners and allied professionals carry 'situated' understandings of the common good that shape the means ostensibly used for determining it. What may at first appear as questions of 'what should be done' are underpinned by conceptions of 'why' something should be done the way it is by whom, when, and where.

This 'substantive' model of understanding how we morally reason from a contingent position within a tradition *of* planning is more modest than the utilitarian and deontologically informed proceduralism or universalism presupposed for identifying the common good in many practitioner activities or the emancipatory aspirations of much planning theory. This is because the interpretive position taken to understanding what the common good may entail evolves from one 'situated' position to another as a tradition extends through reflective debates on distinctions between right/wrong and good/better. Given that such 'situated ethical judgement' (Campbell, 2006) is determined by its qualitative distinctions relative to a potential series of possible alternative evaluations, it is thereby always vulnerable to succession by what is deemed a more appropriate form of substantive reasoning in the future. Hence, Taylor (1995b: 54) when quoting MacIntyre (1977: 455) claims,

> The most that we can claim is that this is the best account which anyone has been able to give so far, and that our beliefs about what the marks of a 'best account so far' are will themselves change in what are at present unpredictable ways.

Accordingly, what the substantive reasoning of situated ethical judgement means is a post-foundational moral epistemology, grounded not in objectivity or postmodern subjectivity but in the 'moral realism' of a limited plurality of ways deemed appropriate for thinking about the common good that are intersubjectively constituted and subject to evolution, yet treated as if they are independent standards. While our views may be vague and change over the course of engagement with others, that does not erase the fact that such changes are interpreted – at least initially – from 'within' the perspectives on the common good we carry into such situations. This is because as planners we are schooled in ideas deemed better/worse through qualitative distinctions on everything from placemaking to public transport. The viewpoints on right/wrong we absorb at university are often finetuned as we proceed from novice to expert in communities of practice that consolidate certain ways of thinking and doing, of deciding and implementing. While the complexities of bureaucratic life and commercial imperatives may mean that the planning activities we undertake do not always wholly harmonise with our moral frameworks, they are nevertheless evaluated relative to the ways of thinking carried by our tradition in the places and junctures in which we find ourselves.

Planning, Planners and the Common Good

How then are we to understand ourselves as planners and our position relative to planning as a discipline? As a philosophical anthropologist of the moral self, Taylor offers some direction on this issue. He draws from Heidegger (1953) on the 'inescapable temporal structure of being in the world' (Taylor, 1989: 47) to envision how people comprehend their situatedness and possibilities in profiling their interpretation of the world and their moral positions regarding it. On the basis that self-understanding has an inherent temporal dimension, Taylor's engagement with Heidegger leads him to conclude that the synthesis of a person's past, present and potential future operates through narrative. Referencing MacIntyre's concept of a 'quest,' Taylor holds that 'just as the self is and must be orientated by a framework which maps moral space, it must also be located in a narrative which tracks its unfolding in time' (Smith, 2002: 97). For as he notes,

> Since we cannot do without an orientation to the good, and since we cannot be indifferent to our place relative to this good, and since this place is something that must always change and become, the issue of the direction of our lives must arise for us. . . . Now we see that this sense of the good has to be woven into my understanding of my life as an unfolding story.
>
> *(Taylor, 1989: 47)*

The issue here is not simply that we interpret our identity as planners relative to what our moral frameworks indicate as the good in a specific context, important though this observation is. Rather, it is that we interpret our position relative to this good as changing over time as we move towards or away from it. Hence, just as an unfolding story appears to have a sense of an ending (Kermode, 1967) that gives it structure and every point therein a position relative to a movement towards an endpoint, so too does the person interpret their current position relative to the good in the context of their past, present and aspired for future as a practitioner or academic. In this way, 'an absolute question about whether we are moving in the right direction always frames our relative questions about specific goods and actions' (Calhoun, 1991: 238). To ignore such a question would be to ignore the whispering of our moral frameworks on what is the right thing to do. Hence, the hypothetical planner we met in Chapter 4 advocating for an amendment to a public bus route and the real Dr Clabby we met in Chapter 5 seeking to 'influence conservation' were motivated by a desire to move closer to what they believed to be the common good. To do otherwise would be a regression, not only in the delivery of the common good but also in the storyline of their self-conception as somebody motivated by a desire to make things better (Lennon, 2015b). Depending on the context and the person living them, such narrative identities may be more implicitly moored to one's sense of integrity than explicitly articulated in public debate. Nonetheless, they enable those involved in this reasoning process to help manage the intersection of the different traditions which crisscross their experience of moral life (planning, parenting, friendship, etc.).

Thus, as was demonstrated by Jane Jacobs (see Chapter 3) and Dr Clabby (see Chapter 5), this 'knowing practice' (Kemmis, 2005) facilitates a conscious reflection on the nature of 'the ends' that planning seeks to achieve. By reflecting on the purpose of what we do in the practice *of* planning, we confront the tradition *of* planning as an activity which seeks to advance the common good. We therein engage with the difficult issue of justifying 'why' we do, or ought to do differently, 'what' we do the 'way' we do it. Such reflection supplies a form of substantive reasoning that practitioners and academics can mine to determine what *should* be done. Indeed, as outlined by Aristotle and drawn upon by MacIntyre, practical and moral reasoning are not discrete forms of reasoning but rather constitute two sides of the same coin. This is because although reasoning is focused on how to achieve a given aim, it is ultimately informed by questions about what that aim should be and how it should be achieved by means of reasoning from within a moral realism that profiles our worldview on what is appropriate. Therefore, the aims that are perceived as choice worthy are identified as such by means of reasoning on what counts as 'worthy'. It is this duality in substantive reasoning that gives shape to MacIntyre's concept of a 'practice'. In the context of planning, one's conception of both the aims to be achieved and the means deemed appropriate to achieving these aims are moulded by one's engagement with the community of practice one seeks out and/or finds oneself immersed within. Thus, substantive reasoning involves a dialectical engagement with the tradition *of* planning. Consequently, substantive reasoning is a form of reasoning informed by, influencing and extending the tradition *of* planning in which it is embedded. Although one's sense of substantive reasoning is influenced by the community of practice within which one is positioned, it deals with managing instances of moral-practical engagement through one's practical activities. Coherence in these activities is provided by channelling and reshaping the tradition *of* planning via our situated ethical judgements as part of a 'quest' to extend that tradition.

Therefore, this book has not sought to proclaim a better means to determine what ought to be done. Instead, it has endeavoured to furnish a more nuanced understanding of how, why and with what effect those involved in planning negotiate the inherent uncertainty, complexity and inevitable normatively of a world brim full of 'wicked problems' in determining what *they* believe ought to be done. While such beliefs may be shaped by the polity and community of practice in which they are embedded, these beliefs are held to be morally 'real' rather than 'relative'. What this book has sought to achieve is thus a refocusing of attention onto an *in situ* conversation with the complexities of action in the here and now, wherein the dialectic of thinking and doing involves context-sensitive reasoning that dynamically reflects and reforms the tradition *of* planning as it engages with difficult questions about determining what ought to be done *in* planning. Accordingly, this approach does not begin with asking 'how do I advance more meaningful participation?,' 'how do I counter-power asymmetries?' or 'how do I promote liberty and equality?' as goals that are 'objectively' right. Instead, it first asks, '*Why* in *this* situation *do I believe* it is best to advance this issue rather than, or in conjunction with another(s)?' In this sense, it reflects Forester's contention that

> an ethically instructive account of planning practice ought to locate that practice in an historical world of influences supporting and threatening the account's notion of good planning. By doing so, the theory can say more than (to a person) 'Do good' or (to a movement) 'Make the revolution.'
>
> *(Forester, 1993: 421)*

The approach advanced in this book is therefore concerned with how practitioners engage with the art of situated ethical judgement as extending the point of planning by confronting live situations through the interpretive tools supplied to them by engagement with their tradition. For as noted by Campbell (2002: 285), 'It is planning's interface with action that gives it its edge; for understanding without implication of action is hollow of meaning, even self-indulgent, while action without understanding is partial.' Seeing the duality of substantive reasoning in the quest for practice excellence suggests that

> ethics is not simply speculative activity, a narrative of thoughts and concepts about hypothetical situations but an explication by doing, a practical application of competences that can be learning only if you are able (and available) to test them in practical experience.
>
> *(Lo Piccolo and Thomas, 2008: 11–12)*

Viewed from the perspective of a narrative extended over time and directed at delivering an ever-evolving concept of the common good, planning is both legitimated and contoured by a desire to do good. To be without such direction would be to question planning's purpose and the function of those theorists and practitioners who dedicate their professional careers to it. As such, planning exists and is propelled by a promise to plan *for* the common good. Anything less undermines its raison d'être.

This brings us full circle to the notion that what unites planning over time and context – its *ipse* identity – is a consistent desire to advance the common good, even if this has been variously interpreted in different ways at different times and in different places. Seen in this context, moral objectivity in planning is impossible. Planners adjudicate implicitly and explicitly; they determine relevant stakeholders, draft policy, set agendas, make decisions and have a large part in profiling new ideas and actions. Even where the modus operandi is ostensibly democratic and experimental, the perspectives of planners on what constitutes democracy and experimentation are central. Planning is at heart a moral discipline whose practitioners are schooled in the art of decision-making. Yes, they learn and use computer software, demographic projections and relevant legal codes. However, these are tools to be mobilised in informing the decision-making process. Technical knowledge may assist substantive reasoning but it does not, nor can it be allowed to, replace it (Arendt, 1964). Legitimised as an activity undertaken in the common good, planning is more than bureaucracy and more than an objective mode of ascertaining preferences. It is the art of situated ethical judgement, learnt, adjusted and applied, wherein 'dealing with the ambiguity of moral dimensions of planning can actually be the moral mandate of planning' (Bickenbach and Hendler, 1994: 175). Positioned in a space of evolving moral questions answered by evolving debates on what *should* be done, planning's relationship with the common good is one where planners attempt to plan *for* the common good. This may be variously understood across time and space but is always directed by a history of substantive reasoning that is (re)interpreted and sometimes transformed.

REFERENCES

Abbey R. (2000) *Charles Taylor*. Teddington, UK: Acumen.
Alexander ER. (2002a) Planning Rights: Towards Normative Criteria for Evaluating Plans. *International Planning Studies* 7: 191–212.
Alexander ER. (2002b) The Public Interest in Planning: From Legitimation to Substantive Plan Evaluation. *Planning Theory* 1: 226–249.
Alexander ER. (2008) Book Review: Thomas L. Harper and Stanley M. Stein, Dialogical Planning in a Fragmented Society: Critically Liberal, Pragmatic, Incremental. *Planning Theory* 7: 108–110.
Alexander ER, Mazza L and Moroni S. (2012) Planning without Plans? Nomocracy or Teleocracy for Socio-Spatial Ordering. *Progress in Planning* 77: 37–87.
Alfasi N. (2009) Planning and the Public Interest: An Editorial Introduction. *Geography Research Forum* 29: 1–6.
Allmendinger P. (2002) *Planning in Postmodern Times*. London, UK: Taylor & Francis.
Allmendinger P. (2017) *Planning Theory*. London, UK: Red Globe Press.
Allmendinger P and Gunder M. (2005) Applying Lacanian Insight and a Dash of Derridean Deconstruction to Planning's 'Dark Side'. *Planning Theory* 4: 87–112.
Allmendinger P and Haughton G. (2012) Post-Political Spatial Planning in England: A Crisis of Consensus? *Transactions of the Institute of British Geographers* 37: 89–103.
Allmendinger P and Haughton G. (2015) Post-Political Regimes in English Planning: From Third Way to Big Society. In: Metzger J, Allmendinger P and Oosterlynck S (eds) *Planning against the Political: Democratic Deficits in European Territorial Governance*. London, UK: Routledge, 29–53.
Amin A and Cirolia LR. (2018) Politics/Matter: Governing Cape Town's Informal Settlements. *Urban Studies* 55: 274–295.
Anscombe GEM. (1958) Modern Moral Philosophy. *Philosophy* 33: 1–19.
Arendt H. (1951) *The Origins of Totalitarianism*. New York: Harcourt Brace Jovanovich.
Arendt H. (1958) *The Human Condition: A Study of the Central Dilemmas Facing Modern Man*. Chicago, IL: University of Chicago Press.
Arendt H. (1964) *Eichmann in Jerusalem: A Report on the Banality of Evil*. New York: Penguin Books Limited.

Argyris C and Schön DA. (1974) *Theory in Practice: Increasing Professional Effectiveness.* San Francisco, CA: Jossey-Bass.
Aristotle. (2014) *Nicomachean Ethics* (translation and commentary by C.D.C. Reeve). Indianapolis, IN: Hackett Publishing Company.
Austin G. (2014) *Green Infrastructure for Landscape Planning: Integrating Human and Natural Systems.* London, UK: Routledge.
Austin JL. (1962) *How To Do Things with Words.* Cambridge, MA: Harvard University Press.
Banfield EC. (1973 (1959)) Ends and Means in Planning. In: Faludi A (ed) *A Reader in Planning Theory.* Oxford, UK: Pergamon, 139–149.
Bassett K. (2014) Rancière, Politics, and the Occupy Movement. *Environment and Planning D: Society and Space* 32: 886–901.
Beauregard RA. (1998) Writing the Planner. *Journal of Planning Education and Research* 18: 93–101.
Beauregard RA. (2015) *Planning Matter: Acting with Things.* Chicago, IL: University of Chicago Press.
Beauregard RA. (2020) *Advanced Introduction to Planning Theory.* Cheltenham, UK: Edward Elgar Publishing.
Benedict M and McMahon E. (2002) Green Infrastructure: Smart Conservation for the 21st Century. *Renewable Resources Journal* 20: 12–17.
Benedict M and McMahon E. (2006) *Green Infrastructure: Linking Landscapes and Communities.* London, UK: Island Press.
Bentham J. (2004) *Utilitarianism and Other Essays.* London, UK: Penguin Books Limited.
Bentham J. (2019) *An Introduction to the Principles of Morals and Legislation.* Dumfries, UK: Anodos Books.
Berger P and Luckmann T. (1966) *The Social Construction of Reality: A Treatise on the Sociology of Knowledge.* London, UK: Penguin Books.
Bernstein R. (1983) *Beyond Objectivism and Relativism: Science, Hermeneutics and Praxis.* Philadelphia, PA: University of Pennsylvania Press.
Bickenbach JE and Hendler S. (1994) The Moral Mandate of the 'Profession' of Planning. In: Thomas H (ed) *Values and Planning.* Abingdon, UK: Ashgate, 162–177.
Bishop R, Phillips J and Yeo WW. (2013) *Postcolonial Urbanism: Southeast Asian Cities and Global Processes.* London, UK: Routledge.
Blackmore C. (2010) *Social Learning Systems and Communities of Practice.* Milton Keynes, UK: Springer.
Blanco H. (1994) *How to Think about Social Problems: American Pragmatism and the Idea of Planning.* Westport, CT: Greenwood Press.
Boelens L. (2010) Theorizing Practice and Practicing Theory: Outlines for an Actor-Relational Approach in Planning. *Planning Theory* 9: 28–62.
Bourdieu P. (1972) *Outline of a Theory of Practice.* Cambridge, UK: Cambridge University Press.
Boyer MC. (1983) *Dreaming the Rational City: The Myth of American City Planning.* Cambridge, MA: MIT Press.
Calhoun C. (1991) Morality, Identity, and Historical Explanation: Charles Taylor on the Sources of the Self. *Sociological Theory* 9: 232–263.
Calhoun C. (2000) *Charles Taylor on Identity and the Social Imaginary.* Unpublish Working Paper. Available at: http://eprints.lse.ac.uk/48046/.
Campbell H. (2002) Planning: An Idea of Value. *The Town Planning Review* 73: 271–288.
Campbell H. (2003) Time, Permanence and Planning: An Exploration of Cultural Attitudes. *Planning Theory & Practice* 4: 461–463.
Campbell H. (2006) Just Planning: The Art of Situated Ethical Judgment. *Journal of Planning Education and Research* 26: 92–106.

Campbell H. (2012a) 'Planning Ethics' and Rediscovering the Idea of Planning. *Planning Theory* 11: 379–399.
Campbell H. (2012b) Planning to Change the World: Between Knowledge and Action Lies Synthesis. *Journal of Planning Education and Research* 32: 135–146.
Campbell H and Marshall R. (1999) Ethical Frameworks and Planning Theory. *International Journal of Urban and Regional Research* 23: 464–478.
Campbell H and Marshall R. (2000) Moral Obligations, Planning, and the Public Interest: A Commentary on Current British Practice. *Environment and Planning B: Planning and Design* 27: 297–312.
Campbell H and Marshall R. (2002) Utilitarianism's Bad Breath? A Re-Evaluation of the Public Interest Justification for Planning. *Planning Theory* 1: 163–187.
Campbell H, Tait M and Watkins C. (2014) Is There Space for Better Planning in a Neoliberal World? Implications for Planning Practice and Theory. *Journal of Planning Education and Research* 34: 45–59.
Certomà C. (2015) Expanding the 'Dark Side of Planning': Governmentality and Biopolitics in Urban Garden Planning. *Planning Theory* 14: 23–43.
Chadwick G. (1971) *A Systems View of Planning: Towards a Theory of Urban and Regional Planning Process*. Oxford, UK: Pergamon.
Chambers SA. (2010) Police and Oligarchy. In: Deranty J-P (ed) *Jacques Rancière: Key Concepts*. Durham, UK: Acumen, 57–68.
Chettiparamb A. (2006a) Fractal Spaces for Planning and Governance. *Town Planning Review* 76: 317–340.
Chettiparamb A. (2006b) Metaphors in Complexity Theory and Planning. *Planning Theory* 5: 71–91.
Chettiparamb A. (2013) Fractal Spatialities. *Environment and Planning. D, Society & Space* 31: 680–692.
Chettiparamb A. (2014) Complexity Theory and Planning: Examining 'Fractals' for Organising Policy Domains in Planning Practice. *Planning Theory* 13: 5–25.
Chettiparamb A. (2016) Articulating 'Public Interest' through Complexity Theory. *Environment and Planning C: Government and Policy* 34: 1284–1305.
Chettiparamb A. (2019) Responding to a Complex World: Explorations in Spatial Planning. *Planning Theory* 18: 429–447.
Clifford B and Tewdwr-Jones M. (2014) *The Collaborating Planner?: Practitioners in the Neoliberal Age*. Bristol, UK: Policy Press.
Comhar. (2010) *Creating Green Infrastructure for Ireland: Enhancing Natural Capital for Human Well Being*. Dublin, Ireland: Comhar SDC.
Connolly J and Steil J. (2009) Introduction: Finding Justice in the City. In: Marcuse P, Connolly J, Novy J, et al. (eds) *Searching for the Just City: Debates in Urban Theory and Practice*. Abingdon, UK: Routledge, 1–15.
Conway D. (1995) *Classical Liberalism: The Unvanquished Ideal*. London, UK: MacMillan Press.
Cooper M and Blair C. (2002) Foucault's Ethics. *Qualitative Inquiry* 8: 511–531.
Cowell R and Lennon M. (2014) The Utilisation of Environmental Knowledge in Land-use Planning: Drawing Lessons for an Ecosystem Services Approach. *Environment and Planning C: Government and Policy* 32: 263–282.
CSO. (2011) *2011 Census of Population: This is Ireland – Highlights from the Census, Part 1*. Available at: www.cso.ie/en/census/census2011reports/census2011thisisirelandpart1/.
Dahl R and Lindblom CE. (1953) *Politics, Economics and Welfare*. New York: Harper & Row.
Davidoff P. (1965) Advocacy and Pluralism in Planning. *Journal of the American Planning Association* 31: 331–338.

Davidoff P and Reiner TA. (1973) A Choice Theory of Planning. In: Faludi A (ed) *A Reader in Planning Theory*. Oxford, UK: Pergamon, 11–39.

Davidson D. (1984) *Inquiries Into Truth and Interpretation: Philosophical Essays Volume 2*. Oxford, UK: Oxford University Press.

Davidson M and Iveson K. (2015) Recovering the Politics of the City: From the 'Post-Political City' to a 'Method of Equality' for Critical Urban Geography. *Progress in Human Geography* 39: 543–559.

Davis O. (2013) *Jacques Rancière*. Cambridge, UK: Polity Press.

Davoudi S. (2015) Planning as Practice of Knowing. *Planning Theory* 14: 316–331.

Davoudi S and Madanipour A. (2013) Localism and Neo-liberal Governmentality. *Town Planning Review* 84: 551–561.

De Neufville JI and Innes JE. (1975) *Social Indicators and Public Policy: Interactive Processes of Design and Application*: Amsterdam and New York: Elsevier Scientific Publishing Company.

d'Entrèves MP and Benhabib S. (1997) *Habermas and the Unfinished Project of Modernity: Critical Essays on The Philosophical Discourse of Modernity*. Cambridge, MA: MIT Press.

de Roo G. (2018) Spatial Planning and the Complexity of Turbulent, Open Environments: About Purposeful Interventions in a World of Non-Linear Change. In: Gunder M, Madanipour A and Watson V (eds) *The Routledge Hanbook of Planning Theory*. Abingdon, UK: Routledge, 314–325.

Dean J. (2001) Publicity's Secret. *Political Theory* 29: 624–650.

Dean M. (2010) *Governmentality: Power and Rule in Modern Society*. London, UK: Sage.

Dear M and Scott AJ. (1981) *Urbanization and Urban Planning in Capitalist Society*. London, UK: Methuen.

Derrida J. (1978) *Writing and Difference*. Chicago, IL: University of Chicago Press.

Dewey J. (1954) *The Public and Its Problems*. Athens, OH: Ohio University Press.

Dewey J. (2008) *The Later Works of John Dewey, 1925–1953: 1938, Logic – The Theory of Inquiry*. Carbondale, IL: Southern Illinois University Press.

Dikeç M. (2005) Space, Politics, and the Political. *Environment and Planning D: Society and Space* 23: 171–188.

Dikec M. (2015) *Space, Politics and Aesthetics*. Edinburgh, Scotland, UK: Edinburgh University Press.

Dikeç M and Swyngedouw E. (2017) Theorizing the Politicizing City. *International Journal of Urban and Regional Research* 41: 1–18.

DoAHGI. (2002) *National Heritage Plan*. Dublin, Ireland: Government of Ireland.

DoECLG. (2013) *Local Area Plans: Guidelines for Planning Authorities*. Dublin, Ireland: Department of the Environment, Community and Local Government (DoECLG).

Dror Y. (1968) *Public Policymaking Reexamined*. San Francisco, CA: Chandler.

Dryzek JS and Niemeyer S. (2006) Reconciling Pluralism and Consensus as Political Ideals. *American Journal of Political Science* 50: 634–649.

Etzioni A. (1967) Mixed-Scanning: A "Third" Approach to Decision-Making. *Public Administration Review* 27: 385–392.

Fainstein NI and Fainstein SS. (1979) New Debates in Urban Planning: The Impact of Marxist Theory within the United States. *International Journal of Urban and Regional Research* 3: 381–403.

Fainstein SS. (2000) New Directions in Planning Theory. *Urban Affairs Review* 35: 451–478.

Fainstein SS. (2005) Planning Theory and the City. *Journal of Planning Education and Research* 25: 121–130.

Fainstein SS. (2009) Planning and the Just City. In: Marcuse P, Connolly J, Novy J, et al. (eds) *Searching for the Just City: Debates in Urban Theory and Practice*. Abingdon, UK: Routledge, 19–39.

Fainstein SS. (2010) *The Just City*. Ithaca, NY: Cornell University Press.
Fainstein SS. (2014) The Just City. *International Journal of Urban Sciences* 18: 1–18.
Fainstein SS. (2016) Spatial Justice and Planning. In: Fainstein SS and DeFilippis J (eds) *Readings in Planning Theory*. Oxford, UK: Wiley, 258–272.
Fainstein SS. (2018) Urban Planning and Social Justice. In: Gunder M, Madanipour A and Watson V (eds) *The Routledge Handbook of Planning Theory*. Abingdon, UK: Routledge, 130–142.
Fainstein SS and DeFilippis J. (2016) *Readings in Planning Theory*. London, UK: Wiley.
Faludi A. (1973) *A Reader in Planning Theory*. Oxford, UK: Pergamon.
Faludi A. (2008 (1973)) The Rationale of Planning Theory. In: Hillier J and Healey P (eds) *Foundations of the Planning Enterprise: Critical Essays in Planning Theory Vol. 1*. Abingdon, UK: Routledge, 375–393.
Fischler R. (2000) Communicative Planning Theory: A Foucauldian Assessment. *Journal of Planning Education and Research* 19: 358–368.
Flathman RE. (1966) *The Public Interest: An Essay Concerning the Normative Discourse of Politics*. New York, New York State: Wiley.
Flynn TR. (2010) *Sartre, Foucault, and Historical Reason, Volume Two: A Poststructuralist Mapping of History*: University of Chicago Press.
Flyvbjerg B. (1996) The Dark Side of Planning: Rationality and 'Realrationalität'. In: Mandelbaum SJ, Mazza L and Burchell RW (eds) *Explorations in Planning Theory*. New Brunswick, NJ: Center for Urban Policy Research Press, 383–394.
Flyvbjerg B. (1998) *Rationality and Power: Democracy in Practice*. London, UK: The University of Chicago Press Ltd.
Flyvbjerg B. (2002) Bringing Power to Planning Research: One Researcher's Praxis Story. *Journal of Planning Education and Research* 21: 353–366.
Flyvbjerg B. (2004) Phronetic Planning Research: Theoretical and Methodological Reflections. *Planning Theory & Practice* 5: 283–306.
Flyvbjerg B and Richardson T. (2002) Planning and Foucault: In Search of the Dark Side of Planning. In: Allmendinger P and Tewdwr-Jones M (eds) *Planning Futures: New Directions for Planning Theory*. Oxford, UK: Routeledge, 44–62.
Forester J. (1982) Planning in the Face of Power. *Journal of the American Planning Association* 48: 67–80.
Forester J. (1989) *Planning in the Face of Power*. Berkeley, CA: University of California Press.
Forester J. (1993) *Critical Theory, Public Policy and Planning Practice*. Albany, NY: State University of New York Press.
Forester J. (1994) Political Judgement and Learning about Value in Transportation Planning: Bridging Habermas and Aristotle. In: Thomas H (ed) *Values and Planning*. Abingdon, UK: Ashgate, 178–204.
Forester J. (1999a) *The Deliberative Practitioner: Encouraging Participatory Planning Processes*. Cambridge, MA: MIT Press.
Forester J. (1999b) Reflections on the Future Understanding of Planning Practice. *International Planning Studies* 4: 175–193.
Forester J. (2009) *Dealing with Differences: Dramas of Mediating Public Disputes*. Oxford, UK: Oxford University Press.
Forester J. (2017) On the Evolution of Critical Pragmatism. In: Haselsberger B (ed) *Encounters In Planning Thought: 16 Autobiographical Essays from Key Thinkers in Spatial Planning*. London, UK: Routledge, 280–296.
Forrester J. (1969) *Urban Dynamics*. Cambridge, MA: MIT Press.
Foucault M. (1977) *Power/Knowledge: Selected Interviews and Other Writings, 1972–1977*. New York: Randon House Inc.

Foucault M. (1982) The Subject of Power. *Critical Inquiry* 8: 777–795.
Foucault M. (1984a) On the Geneology of Ethics: An Overview of Work in Progress. In: Rabinow P (ed) *The Foucault Reader*. New York: Pantheon.
Foucault M. (1984b) *The Use of Pleasure: Volume II of the History of Sexuality*. New York: Vintage.
Foucault M. (1986) Kant on Enlightenment and Revolution. *Economy and Society* 15: 88–96.
Foucault M. (1987) The Ethic of Care for the Self as a Practice of Freedom: An Interview with Michel Foucault on January 20, 1984. *Philosophy & Social Criticism* 12: 112–131.
Foucault M. (1990) *The History of Sexuality, Volume I: An Introduction*. New York: Vintage Press.
Foucault M. (1991) *Discipline and Punishment*. Harmonsworth, UK: Penguin.
Foucault M. (2002) *The Order of Things: An Archaeology of the Human Sciences*. Abingdon, UK: Routledge.
Fox-Rogers L. (2019) The Dark Side of Community: Clientalism, Corruption and Legitimacy in Rural Planning. In: Scott M, Gallent N and Gkartzios M (eds) *The Routledge Companion to Rural Planning*. Abingdon, UK: Routledge, 142–151.
Fox-Rogers L and Murphy E. (2014) Informal Strategies of Power in the Local Planning System. *Planning Theory* 13: 244–268.
Fox-Rogers L and Murphy E. (2016) Self-perceptions of the Role of the Planner. *Environment and Planning B: Planning and Design* 43: 74–92.
Frankfurt H. (1971) Freedom of the Will and the Concept of a Person. *Journal of Philosophy* 67: 5–20.
Fraser N and Honneth A. (2003) *Redistribution or Recognition?: A Political-Philosophical Exchange*. London, UK: Verso Press.
Frazer A and Lacey N. (1993) *The Politics of Community: A Feminist Critique of the Liberal-Communitarian Debate*. Toronto, Ontario: University of Toronto Press.
Friedmann J. (1973) *Retracking America: A Theory of Transactive Planning*. Harlow, UK: Anchor Press.
Friedmann J. (1987) *Planning in the Public Domain: From Knowledge to Action*. Princeton, NJ: Princeton University Press.
Friedmann J. (2011) *Insurgencies: Essays in Planning Theory*. Abingdon, UK: Routledge.
Gans H. (1968) *People and Plans*. New York: Basic Books.
Giddens A. (1984) *The Constitution of Society: Outline of the Theory of Structuration*. Cambridge, UK: Polity Press.
Giddens A. (1990) *The Consequencies of Modernity*. Cambridge, UK: Polity.
Gramsci A. (1971) *Selections from the Prison Notebooks of Antonio Gramsci*. New York: International Publishers.
Grant J. (1994) On Some Public Uses of Planning 'Theory': Rhetoric and Expertise in Community Planning Disputes. *Town Planning Review* 65: 59–78.
Grant J. (2005) Rethinking the Public Interest as a Planning Concept. *Plan Canada* 45: 48–50.
Gualini E. (2015) *Planning and Conflict: Critical Perspectives on Contentious Urban Developments*. Abingdon, UK: Routledge.
Gualini E, Allegra M and Mourato JM. (2015) *Conflict in the City: Contested Urban Spaces and Local Democracy*. Berlin, Germany: Jovis.
Gunder M. (2004) Shaping the Planner's Ego-Ideal: A Lacanian Interpretation of Planning Education. *Journal of Planning Education and Research* 23: 299–311.
Gunder M. (2014) Fantasy in Planning Organisations and their Agency: The Promise of Being at Home in the World. *Urban Policy and Research* 32: 1–15.
Gunder M. (2016) Planning's 'Failure' to Ensure Efficient Market Delivery: A Lacanian Deconstruction of this Neoliberal Scapegoating Fantasy. *European Planning Studies* 24: 21–38.

Gunder M and Hillier J. (2004) Conforming to the Expectations of the Profession: A Lacanian Perspective on Planning Practice, Norms and Values. *Planning Theory & Practice* 5: 217–235.

Gunder M and Hillier J. (2009) *Planning in Ten Words Or Less: A Lacanian Entanglement with Spatial Planning*. Farnham, Surrey: Ashgate.

Gunder M, Madanipour A and Watson V. (2018) *The Routledge Handbook of Planning Theory*. Abingdon, UK: Routledge.

Habermas J. (1970) Towards a Theory of Communicative Competence. *Inquiry* 13: 360–375.

Habermas J. (1971) *Toward a Rational Society*. Portsmouth, New Hampshire: Heinemann Educational Books.

Habermas J. (1979) *Communication and the Evolution of Society*. Portsmouth, New Hampshire: Heinemann Educational Books.

Habermas J. (1985) *The Theory of Communicative Action Vol. 1: Reason and the Rationalization of Society*. Boston, MA: Beacon Press.

Habermas J. (1987) *The Theory of Communicative Action Vol. 2: System and Lifeworlds: A Critique of Functionalist Reason*. Boston, MA: Beacon Press.

Hall P. (2014) *Cities of Tomorrow: An Intellectual History of Urban Planning and Design Since 1880*. Chichester, UK: Wiley Blackwell.

Harper TL and Stein SM. (1992) The Centrality of Normative Ethical Theory to Contemporary Planning Theory. *Journal of Planning Education and Research* 11: 105–116.

Harper TL and Stein SM. (1996) A Classical Liberal (Libertarian) Approach to Planning Theory. In: Mendelbaum S, Mazza L and Burchell RW (eds) *Explorations in Planning Theory*. New Brunswick, NJ: Centre for Urban Policy Research, 11–29.

Harper TL and Stein SM. (2006) *Dialogical Planning in a Fragmented Society: Critically Liberal, Pragmatic, Incremental*. Piscataway, NJ: Transaction Publishers.

Harris N. (2002) Collaborative Planning: From Theoretical Foundations to Practice Forms. In: Allmendinger P and Tewdwr-Jones M (eds) *Planning Futures: New Directions for Planning Theory*. Abingdon, UK: Routledge.

Harris N. (2011) Discipline, Surveillance, Control: A Foucaultian Perspective on the Enforcement of Planning Regulations. *Planning Theory & Practice* 12: 57–76.

Hartmann T. (2012) Wicked Problems and Clumsy Solutions: Planning as Expectation Management. *Planning Theory* 11: 242–256.

Harvey D. (1973) *Social Justice and the City*. London, UK: Edward Arnold.

Harvey D. (1978) On Planning the Ideology of Planning. In: Beauregard RA and Sternlieb G (eds) *Planning Theory in the 1980s: A Search for Future Directions*. New Brunswick, NJ: Rutgers University Press, 213–225.

Hayek FA. (1978) *New Studies in Philosophy, Politics, Economics and the History of Ideas*. London: Routledge & Kegan-Paul.

Hayek FA. (1982) *Law, Legislation and Liberty*. London: Routledge.

Healey P. (1992a) A Planner's Day: Knowledge and Action in Communicative Practice. *Journal of the American Planning Association* 58: 9–20.

Healey P. (1992b) Planning through Debate: The Communicative Turn in Planning Theory. *Town Planning Review* 63: 143–162.

Healey P. (1996) The Communicative Work of Development Plans. In: Mendelbaum S, Mazza L and Burchell RW (eds) *Explorations in Planning Theory*. New Brunswick, NJ: Centre for Urban Policy Research, 263–288.

Healey P. (1997) *Collaborative Planning: Shaping Places in Fragmented Societies*. Basingstoke, UK: Palgrave Macmillan.

Healey P. (1998) Collaborative Planning in a Stakeholder Society. *Town Planning Review* 69: 1–21.

Healey P. (1999) Institutionalist Analysis, Communicative Planning, and Shaping Places. *Journal of Planning Education and Research* 19: 111–121.
Healey P. (2003) Collaborative Planning in Perspective. *Planning Theory* 2: 101–123.
Healey P. (2006) *Collaborative Planning: Shaping Places in Fragmented Societies*. Basingstoke, UK: Palgrave Macmillan.
Healey P. (2009) The Pragmatic Tradition in Planning Thought. *Journal of Planning Education and Research* 28: 277–292.
Healey P. (2012a) Communicative Planning: Practices, Concepts, and Rhetorics. In: Sanyal B, Vale LJ and Rosan CD (eds) *Planning Ideas That Matter: Livability, Territoriality, Governance, and Reflective Practice*. Cambridge, MA: The MIT Press, 333–357.
Healey P. (2012b) Performing Place Governance Collaboratively: Planning as a Communicative Process. In: Fischer F and Gottweis H (eds) *The Argumentative Turn Revisited: Public Policy as Communicative Practice*. London, UK: Duke University Press.
Healey P. (2017) Finding My Way: A Life of Inquiry into Planning Urban Development Processes and Place Governance. In: Haselsberger B (ed) *Encounters In Planning Thought: 16 Autobiographical Essays from Key Thinkers in Spatial Planning*. London, UK: Routledge, 107–125.
Healey P and Hillier J. (2008) Introduction to Part III. In: Hillier J and Healey P (eds) *Foundations of the Planning Enterprise: Critical Essays in Planning Theory Vol. 1*. Aldershot, UK: Ashgate, 299–305.
Healey P and Hillier J. (2016) *The Ashgate Research Companion to Planning Theory: Conceptual Challenges for Spatial Planning*. Abingdon, UK: Routledge.
Heidegger M. (1953) *Being and Time*. Tubingen, Germany: Max Niemeyer Verlag.
Held D. (1970) *The Public Interest and Individual Interests*. New York: Basic Books.
Hendler S. (1995) *Planning Ethics: A Reader in Planning Theory, Practice, and Education*. Piscataway, NJ: Transaction Publications.
Higgins C. (2003) MacIntyre's Moral Theory and the Possibility of an Aretaic Ethics of Teaching. *Journal of Philosophy of Education* 37: 279–292.
Hillier J. (2003) Agonizing over Consensus: Why Habermasian Ideals Cannot be Real. *Planning Theory* 2: 37–59.
Hillier J. (2011) Strategic Navigation across Multiple Planes: Towards a Deleuzean-inspired Methodology for Strategic Spatial Planning. *Town Planning Review* 82: 503–527.
Hillier J. (2015) If Schrödinger's Cat Miaows in the Suburbs, Will Anyone Hear? *Planning Theory* 14: 425–443.
Hillier J. (2018) Lines of Becoming. In: Gunder M, Madanipour A and Watson V (eds) *The Routledge Handbook of Planning Theory*. Abingdon, UK: Routledge, 337–350.
Hillier J and Healey P. (2008a) *Foundations of the Planning Enterprise: Critical Essays in Planning Theory Vol. 1*. Aldershot, UK: Ashgate.
Hillier J and Healey P. (2008b) Introduction to Part I. In: Hillier J and Healey P (eds) *Contemporary Movements in Planning Theory: Critical Essays in Planning Theory Vol. 3*. Abingdon, UK: Routledge, 3–10.
Hirt S and Zahm DL. (2012) *The Urban Wisdom of Jane Jacobs*. New York: Routledge.
Hoch C. (1984) Doing Good and Being Right. *Journal of the American Planning Association* 50: 335–345.
Hoch C. (1993) Commentary. *Journal of Planning Education and Research* 12: 93–95.
Hoch C. (1996) A Pragmatic Inquiry About Planning and Power. In: Mendelbaum S, Mazza L and Burchell RW (eds) *Explorations in Planning Theory*. Piscataway, NJ: Transaction Publishers, 30–44.
Hoch C. (2018) Neo-Pragmatist Planning Theory. In: Gunder M, Madanipour A and Watson V (eds) *The Routledge Handbook of Planning Theory*. Abingdon, UK: Routledge, 118–129.

Hoch C. (2019) *Pragmatic Spatial Planning: Practial Theory for Professionals*. Abingdon, UK: Routledge.

Honneth A and Rancière J. (2016) *Recognition or Disagreement: A Critical Encounter on the Politics of Freedom, Equality, and Identity*. New York: Columbia University Press.

Howe E. (1992) Professional Roles and the Public Interest in Planning. *Journal of Planning Literature* 6: 230–248.

Howe E. (1994) *Acting on Ethics in City Planning*. New Brunswick, NJ: Center for Urban Policy Research.

Howe E and Kaufman J. (1979) The Ethics of Contemporary American Planners. *Journal of the American Planning Association* 45: 243–255.

Huxley M. (1994) Panoptica: Utilitarianism and Land-use Control. In: Gibson K and Watson S (eds) *Metropolis Now: Planning and the Urban in Contemporary Australia*. Sydney, Australia: Pluto Press, 148–160.

Huxley M. (2000) The Limits to Communicative Planning. *Journal of Planning Education and Research* 19: 369–377.

Huxley M. (2006) Spatial Rationalities: Order, Environment, Evolution and Government. *Social & Cultural Geography* 7: 771–787.

Huxley M. (2018) Countering 'The Dark Side' of Planning: Power, Governmentality, Counter-Conduct. In: Gunder M, Madanipour A and Watson V (eds) *The Routledge Handbook of Planning Theory*. Abingdon, UK: Routledge, 207–220.

Inch A. (2010) Culture Change as Identity Regulation: The Micro-Politics of Producing Spatial Planners in England. *Planning Theory & Practice* 11: 359–374.

Inch A. (2012) 'Cultural Work', Spatial Planning and the Politics of Renewing Public Sector Planning Professionalism in England. *The Town Planning Review* 83: 513–532.

Inch A. (2018) 'Cultural Work' and the Remaking of Planning's 'Apparatus of Truth'. In: Gunder M, Madanipour A and Watson V (eds) *The Routledge Handbook of Planning Theory*. Abingdon, UK: Routledge, 194–206.

Innes JE. (1996) Planning through Consensus Building: A New View of the Comprehensive Planning Ideal. *Journal of the American Planning Association* 62: 460–472.

Innes JE and Booher DE. (2003) Collaborative Policymaking: Governance through Dialogue. In: Hajer M and Wagenaar H (eds) *Deliberative Policy Analysis: Understanding Governance in the Network Society*. Cambridge, UK: Cambridge University Press.

Innes JE and Booher DE. (2010) *Planning with Complexity: An Introduction to Collaborative Rationality for Public Policy*. New York: Taylor & Francis.

Innes JE and Booher DE. (2015) A Turning Point for Planning Theory? Overcoming Dividing Discourses. *Planning Theory* 14: 195–213.

Jackson S, Porter L and Johnson LC. (2017) *Planning in Indigenous Australia: From Imperial Foundations to Postcolonial Futures*. Abingdon, UK: Routledge.

Jacobs J. (1961) *The Death and Life of Great American Cities*. New York: Penguin.

James I. (2014) *The New French Philosophy*. Cambridge, UK: Polity Press.

Jepson P and Ladle R. (2010) *Conservation: A Beginner's Guide*. Oxford, UK: Oneworld Publications.

Kabisch N, Korn H, Stadler J, et al. (2017) *Nature-Based Solutions to Climate Change Adaptation in Urban Areas: Linkages between Science, Policy and Practice*. Cham, Switzerland: Springer International Publishing.

Kalberg S. (1980) Max Weber's Types of Rationality: Cornerstones for the Analysis of Rationalization Processes in History. *American Journal of Sociology* 85: 1145–1179.

Kant I. ((1788) 2012) *The Critique of Practical Reason*. San Antonio, TX: Bibliotech Press.

Kaufman JL. (1981) Ethics and Planning: Some Insights from the Outside. *Journal of American Planning Association* 47: 196–199.

Kemmis S. (2005) Knowing Practice: Searching for Saliences. *Pedagogy, Culture & Society* 13: 391–426.
Kermode F. (1967) *The Sense of an Ending: Studies in the Theory of Fiction.* New York: Oxford University Press.
Klosterman RE. (1985) Arguments for and against Planning. *Town Planning Review* 1: 5–20.
Knight K. (2007) *Aristotelian Philosophy: Ethics and Politics from Aristotle to MacIntyre.* Cambridge, UK: Polity Press.
Krumholz N, Cogger JM and Liner JH. (1975) The Cleveland Policy Planning Report. *Journal of the American Institute of Planners* 41: 298–304.
Kühn M. (2020) Agonistic Planning Theory Revisited: The Planner's Role in Dealing with Conflict. *Planning Theory* [Online]. https://doi.org/1473095220953201.
Laclau E. (1990) *New Reflections on the Revolution of Our Time.* London, UK: Verso.
Laclau E and Mouffe C. (1985) *Hegemony and Socialist Strategy.* London, UK: Verson.
Lauria M and Long MF. (2017) Planning Experience and Planners' Ethics. *Journal of the American Planning Association* 83: 202–220.
Lauria M and Long MF. (2019) Ethical Dilemmas in Professional Planning Practice in the United States. *Journal of the American Planning Association* 85: 393–404.
Lave J and Wenger E. (1991) *Situated Learning: Legitimate Peripheral Participation.* Cambridge, UK: Cambridge University Press.
Legacy C. (2016) Transforming Transport Planning in the Postpolitical Era. *Urban Studies* 53: 3108–3124.
Legacy C, Metzger J, Steele W, et al. (2019) Beyond the Post-political: Exploring the Relational and Situated Dynamics of Consensus and Conflict in Planning. *Planning Theory* 18: 273–281.
Lennon M. (2014) Presentation and Persuasion: The Meaning of Evidence in Irish Green Infrastructure Policy. *Evidence & Policy: A Journal of Research, Debate and Practice* 10: 167–186.
Lennon M. (2015a) Explaining the Currency of Novel Policy Concepts: Learning from Green Infrastructure Planning. *Environment and Planning C: Government and Policy* 33: 1039–1057.
Lennon M. (2015b) Finding Purpose in Planning. *Journal of Planning Education and Research* 35: 63–75.
Lennon M. (2015c) Green Infrastructure and Planning Policy: A Critical Assessment. *Local Environment* 20: 957–980.
Lennon M. (2017) On 'the Subject' of Planning's Public Interest. *Planning Theory* 16: 150–168.
Lennon M. (2018) Grasping Green Infrastructure: An Introduction to the Theory and Practice of a Diverse Environmental Planning Approach. In: Davoudi S, Blanco H, Cowell R, et al. (eds) *Routledge Companion to Environmental Planning and Sustainability.* London, UK: Routledge, 277–288.
Lennon M. (2019) Grasping Green Infrastructure: An Introduction to the Theory and Practice of a Diverse Environmental Planning Approach. In: Davoudi S, Cowell R, White I, et al. (eds) *The Routledge Companion to Environmental Planning.* Abingdon, UK: Routledge, 277–289.
Lennon M and Fox-Rogers L. (2017) Morality, Power and the Planning Subject. *Planning Theory* 16: 364–383.
Lennon M and Moore D. (2019) Planning, 'Politics' and the Production of Space: The Formulation and Application of a Framework for Examining the Micropolitics of Community Place-making. *Journal of Environmental Policy & Planning* 21: 117–133.
Lennon M, Scott M, Collier M, et al. (2016) The Emergence of Green Infrastructure as Promoting the Centralisation of a Landscape Perspective in Spatial Planning – The Case of Ireland. *Landscape Research* 42: 146–163.

Lennon M and Waldron R. (2019) De-democratising the Irish Planning System. *European Planning Studies* 28: 1607–1625.

Lindblom CE. (1959) The Science of "Muddling Through". *Public Administration Review* 19: 78–88.

Lo Piccolo F and Thomas H. (2008) Research Ethics in Planning: A Framework for Discussion. *Planning Theory* 7: 7–23.

Loh CG and Arroyo RL. (2017) Special Ethical Considerations for Planners in Private Practice. *Journal of the American Planning Association* 83: 168–179.

Loh CG and Norton RK. (2013) Planning Consultants and Local Planning: Roles and Values. *Journal of the American Planning Association* 79: 138–147.

Low NP. (1991a) *Planning, Politics & State: Political Foundations of Planning Thought*. London, UK: Unwin Hyman Ltd.

Low NP. (1991b) *Planning, Politics and the State: Political Foundations of Planning Thought*. Oxford, UK: Unwin Hyman Ltd.

Lutz CS. (2009) *Tradition in the Ethics of Alasdair MacIntyre: Relativism, Thomism, and Philosophy*. New York: Lexington Books.

Lutz CS. (2012) *Reading Alasdair MacIntyre's After Virtue*. London, UK: Continuum International Publishing.

MacDonald KM. (1995) *The Sociology of the Professions*. London, UK: Sage.

MacIntyre AC. (1977) Epistemological Crises, Dramatic Narrative and the Philosophy of Science. *The Monist* 60: 453–472.

MacIntyre AC. (1984) *After Virtue: A Study in Moral Theory*. Notre Dame, IN: University of Notre Dame Press.

MacIntyre AC. (1988) *Whose Justice? Which Rationality?*. Notre Dame, IN: University of Notre Dame Press.

MacIntyre AC. (1990) *Three Rival Versions of Moral Enquiry: Encyclopedia, Genealogy, and Tradition*. Notre Dame, IN: University of Notre Dame Press.

MacIntyre AC. (2007) *After Virtue: A Study in Moral Theory*. Notre Dame, IN: University of Notre Dame Press.

Madanipour A. (2015) *Planning Theory*. Abingdon, UK: Routledge.

Maidment C. (2016) In the Public Interest? Planning in the Peak District National Park. *Planning Theory* 15: 366–385.

Mandelbaum SJ. (2000) *Open Moral Communities*. Cambridge, MA: MIT Press.

Mandelbaum SJ, Mazza L and Burchell RW. (1996) *Explorations in Planning Theory*. New Brunswick, NJ: Centre for Urban Policy Research.

Mannheim K. (1940) *Man and Society in an Age of Reconstruction*. London, UK: Routledge.

Marcuse P, Connolly J, Novy J, et al. (2009) *Searching for the Just City: Debates in Urban Theory and Practice*. London: Routledge.

Margolis J. (1998) Objectivity as a Problem: An Attempt at an Overview. *Annals of the American Academy of Political and Social Science* 560: 55–68.

Mattila H. (2016) Can Collaborative Planning go Beyond locally Focused Notions of the "Public Interest"? The Potential of Habermas' Concept of "Generalizable Interest" in Pluralist and Trans-Scalar Planning Discourses. *Planning Theory* 15: 344–365.

May T. (2008) *The Political Thought of Jacques Rancière: Creating Equality*. Edinburgh, Scotland, UK: Edinburgh University Press.

May T. (2010) Wrong, Disagreement, Subjectification. In: Deranty J-P (ed) *Jacques Rancière: Key Concepts*. Durham, UK: Acumen, 69–79.

McAuliffe C and Rogers D. (2019) The Politics of Value in Urban Development: Valuing Conflict in Agonistic Pluralism. *Planning Theory* 18: 300–318.

McClymont K. (2011) Revitalising the Political: Development Control and Agonism in Planning Practice. *Planning Theory* 10: 239–256.
McClymont K. (2019) Articulating Virtue: Planning Ethics within and Beyond Post Politics. *Planning Theory* 18: 282–299.
McKay S, Murray M and Macintyre S. (2012) Justice as Fairness in Planning Policy-Making. *International Planning Studies* 17: 147–162.
McLoughlin JB. (1969) *Urban and Regional Planning: A Systems Approach*. London, UK: Faber.
Mell I. (2016) *Global Green Infrastructure: Lessons for Successful Policy-making, Investment and Management*. London, UK: Taylor & Francis.
Mell I. (2019) *Green Infrastructure Planning: Reintegrating Landscape in Urban Planning*. London, UK: Lund Humphries.
Merchart O. (2007) *Post-Foundational Political Thought*. Edinburgh, Scotland, UK: Edinburgh University Press.
Merleau-Ponty M. (2002) *Phenomenology of Perception*. Abingdon, UK: Routledge.
Meth P. (2010) Unsettling Insurgency: Reflections on Women's Insurgent Practices in South Africa. *Planning Theory & Practice* 11: 241–263.
Metzger J. (2018) Postpolitics and Planning. In: Gunder M, Madanipour A and Watson V (eds) *The Routledge Handbook of Planning Theory*. Abingdon, UK: Routledge, 180–193.
Metzger J, Allmendinger P and Oosterlynck S. (2015) The Contested Terrain of European Territorial Governance: New Perspectives on Democratic Deficits and Political Displacements. In: Metzger J, Allmendinger P and Oosterlynck S (eds) *Planning against the Political*. Abingdon, UK: Routledge, 1–28.
Meyerson M and Banfield EC. (1955) *Politics, Planning and the Public Interest*. New York: Free Press.
Mill JS. (2015) *On Liberty, Utilitarianism and Other Essays*. Oxford, UK: Oxford University Press.
Miller D. (1994) Virtues, Practices and Justice. In: Horton J and Mendus S (eds) *After MacIntyre: Critical Perspectives on the Work of Alasdair MacIntyre*. Notre Dame, IN: University of Notre Dame Press, 245–264.
Mills S. (2003) *Michel Foucault*. Oxford, UK: Routledge.
Miraftab F. (2009) Insurgent Planning: Situating Radical Planning in the Global South. *Planning Theory* 8: 32–50.
Moroni S. (2004) Towards a Reconstruction of the Public Interest Criterion. *Planning Theory* 3: 151–171.
Moroni S. (2010) Rethinking the Theory and Practice of Land-use Regulation: Towards Nomocracy. *Planning Theory* 9: 137–155.
Moroni S. (2014) Towards a General Theory of Contractual Communities. In: Andersson D and Moroni S (eds) *Cities and Private Planning*. Cheltenham: Edward Elgar, 38–65.
Moroni S. (2018) The Public Interest. In: Gunder M, Madanipour A and Watson V (eds) *The Routledge Handbook of Planning Theory*. Abingdon, UK: Routledge, 69–80.
Moroni S. (2019) Constitutional and Post-constitutional Problems: Reconsidering the Issues of Public Interest, Agonistic Pluralism and Private Property in Planning. *Planning Theory* 18: 5–23.
Mosteller T. (2008) *Relativism in Contemporary American Philosophy: MacIntyre, Putnam, and Rorty*. London, UK: Bloomsbury Academic.
Mouat C, Legacy C and March A. (2013) The Problem is the Solution: Testing Agonistic Theory's Potential to Recast Intractable Planning Disputes. *Urban Policy and Research* 31: 150–166.
Mouffe C. (1999) Deliberative Democracy or Agonistic Pluralism? *Social research* 66: 745–758.

Mouffe C. (2000) *The Democratic Paradox*. London, UK: Verso.
Mouffe C. (2005) *On the Political*. Abingdon, UK: Routledge.
Mouffe C. (2013) *Agonistics: Thinking the World Politically*. London, UK: Verso.
Mulhall S and Swift A. (1996) *Liberals and Communitarians*. Oxford, UK: Blackwell.
Murphy E and Fox-Rogers L. (2015) Perceptions of the Common Good in Planning. *Cities* 42, Part B: 231–241.
Murphy MC. (2003) MacIntyre's Political Philosophy. In: Murphy MC (ed) *Alasdair MacIntyre*. Cambridge, UK: Cambridge University Press.
Nagel T. (1986) *The View from Nowhere*. Oxford, UK: Oxford University Press.
Nicholas JL. (2012) *Reason, Tradition and The Good*. Notre Dame, IN: University of Notre Dame Press.
Nietzsche F. (2009 (1872)) Homer's Contest. In: Pearson KA and Large D (eds) *The Nietzsche Reader*. Oxford, UK: Wiley, 95–100.
Nussbaum M. (2008) Female Capabilities, Female Human Beings. In: Pogge T and Mollendorf D (eds) *Global Justice; Seminal Essays*. Paragon House Publishers, 495–551.
O'Callaghan C, Boyle M and Kitchin R. (2014) Post-politics, Crisis, and Ireland's 'Ghost Estates'. *Political Geography* 42: 121–133.
Othengrafen F and Levin-Keitel M. (2019) Planners between the Chairs: How Planners (Do Not) Adapt to Transformative Practices. *Urban Planning* 4: 111–138.
Owens S and Cowell R. (2011) *Land and Limits: Interpreting Sustainability in the Planning Process*. New York: Routledge.
Pearlmutter D, Calfapietra C, Samson R, et al. (2017) *The Urban Forest: Cultivating Green Infrastructure for People and the Environment*. Cham, Switzerland: Springer International Publishing.
Peirce CS. (1955) *Philosophical Writings of Peirce*. New York: Dover Publications.
Pinkard T. (2003) MacIntyre's Critique of Modernity. In: Murphy MC (ed) *Alasdair MacIntyre*. Cambridge, UK: Cambridge University Press, 176–200.
Pitkin HF. (1972) *Wittgenstein and Justice: On the Significance of Ludwig Wittgenstein for Social and Political Thought*. Berkeley, CA: University of California Press.
Pløger J. (2004) Strife: Urban Planning and Agonism. *Planning Theory* 3: 71–92.
Pløger J. (2008) Foucault's Dispositif and the City. *Planning Theory* 7: 51–70.
Pløger J. (2018) Conflict and Agonism. In: Gunder M, Madanipour A and Watson V (eds) *The Routledge Handbook of Planning Theory*. Abingdon, UK: Routledge, 264–275.
Popper K. (1959) *The Logic of Scientific Discovery*. London, UK: Hutchingson.
Porter J. (2003) Tradition in the Recent Work of Alasdair MacIntyre. In: Murphy MC (ed) *Alasdair MacIntyre*. Cambridge, UK: Cambridge University Press, 38–69.
Porter L. (2016) *Unlearning the Colonial Cultures of Planning*. Abingdon, UK: Routledge.
Powell WW and DiMaggio PJ. (1991) *The New Institutionalism in Organisational Analysis*. Chicago, USA: University of Chicago Press.
Putnam H. (1981) *Reason, Truth and History*. Cambridge, UK: Cambridge University Press.
Puustinen S, Mäntysalo R and Jarenko K. (2017) The Varying Interpretations of Public Interest: Making Sense of Finnish Urban Planners' Conceptions. *Current Urban Studies* 5: 82–96.
Quine WVO. (1969) *Ontological Relativity and Other Essays*. London, UK: Columbria University Press.
Raco M. (2014) The Post-politics of Sustainability Planning: Privatisation and the Demise of Democratic Government. In: Swyngedouw E and Wilson J (eds) *The Post-Political and Its Discontents: Spaces of Depoliticization, Spectres of Radical Politics*. Edinburgh, Scotland, UK: Edinburgh University Press, 25–47.
Raco M and Imrie R. (2000) Governmentality and Rights and Responsibilities in Urban Policy. *Environment and Planning. A* 32: 2187–2204.

Rancière J. (1999) *Disagreement: Politics and Philosophy*. Minneapolis, MN: University of Minnesota Press.
Rancière J. (2000) Dissenting Words: A Conversation with Jacques Rancière, D. Panagia (trans.). *Diacritics* 30: 113–126.
Rancière J. (2004a) Introducing Disagreement. *Angelaki* 9: 3–9.
Rancière J. (2004b) *The Politics of Aesthetics: The Distribution of the Sensible*. London, UK: Continuum.
Ratcliffe J. (1981) *An Introduction to Town and Country Planning*. London, UK: Hutchinson.
Rawls J. (1971) *A Theory of Justice*. Cambridge, MA: Harvard University Press.
Reade E. (1997) Planning in the Future or Planning of the Future. In: Blowers A and Evans B (eds) *Town Planning in the 21st Century*. London, UK: Routledge, 71–103.
Richards MA. (2018) *Regreening the Built Environment*. Abingdon, UK: Routledge.
Ricoeur P. (1970) *Freud and Philosophy: An Essay on Interpretation*. London, UK: Yale University Press.
Ricoeur P. (1995) *Oneself as Another*. Chicago, IL: University of Chicago Press.
Rittel HW and Webber MM. (1973) Dilemmas in a General Theory of Planning. *Policy Sciences* 4: 155–169.
Rorty R. (1991) *Objectivity, Relativism and Truth: Philosophical Papers Vol. 1*. Cambridge, UK: Cambridge University Press.
Rose N and Miller P. (1992) Political Power beyond the State: Problematics of Government. *British Journal of Sociology* 43: 173–205.
Rouse DC and Bunster-Ossa IF. (2013) *Green Infrastructure: A Landscape Approach*. Washington, DC: American Planning Association.
Roweis S. (1983) Urban Planning as Professional Mediation of Territorial Politics. *Environment and Planning D: Society and Space* 1: 139–162.
Roy A. (2009) Why India Cannot Plan Its Cities: Informality, Insurgence and the Idiom of Urbanization. *Planning Theory* 8: 76–87.
RTPI. (2019) *Corporate Strategy 2020–2030*. London, UK: Royal Town Planning Institute (RTPI).
Rydin Y and Tate L. (2016) *Actor Networks of Planning: Exploring the Influence of Actor Network Theory*. Abingdon, UK: Routledge.
Sager T. (1994) *Communicative Planning Theory*. Aldershot, UK: Avebury.
Sager T. (2012) *Reviving Critical Planning Theory: Dealing with Pressure, Neo-liberalism, and Responsibility in Communicative Planning*. Abingdon, UK: Routledge.
Sager T. (2018) Communicative Planning. In: Gunder M, Madanipour A and Watson V (eds) *The Routledge Handbook of Planning Theory*. Abingdon, UK: Routledge, 93–104.
Salet W. (2018a) *Public Norms and Aspirations: The Turn to Institutions in Action*. Abingdon, UK: Routledge.
Salet W. (2018b) *The Routledge Handbook of Institutions and Planning in Action*. Abingdon, UK: Routledge.
Sandel MJ. (1982) *Liberalism and the Limits of Justice*. Cambridge, UK: Cambridge University Press.
Sandel MJ. (1984) The Procedural Republic and the Unencumbered Self. *Political Theory* 12: 81–96.
Sandercock L. (1995) Voices from the Borderlands: A Meditation on a Metaphor. *Journal of Planning Education and Research* 14: 77–88.
Sandercock L. (1998) *Towards Cosmopolis: Planning for Multicultural Cities*. London, UK: Wiley Blackwell.
Sandercock L. (2004) Towards a Planning Imagination for the 21st Century. *Journal of the American Planning Association* 70: 133–141.

Sandercock L and Dovey K. (2002) Pleasure, Politics, and the "Public Interest": Melbourne's Riverscape Revitalization. *Journal of the American Planning Association* 68: 151–164.

Sarbib JL. (1983) The University of Chicago Program in Planning. *Journal of Planning Education and Research* 2: 77–81.

Sartre JP. (1969) *Being and Nothingness: An Essay on Phenomenological Ontology.* Abingdon, UK: Routledge.

Schmitt C. (2008 [1932]) *The Concept of the Political: Expanded Edition.* Chicago, IL: University of Chicago Press.

Schön DA. (1971) *Beyond the Stable State.* London, UK: Temple Smith.

Schön DA. (1991) *The Reflective Practitioner: How Professionals Think in Action.* London, UK: Ashgate Publishing Limited.

Schön DA and Rein M. (1994) *Frame Reflection: Towards the Resolution of Intractable Policy Controversies.* New York: Basic Books.

Schoneboom A, Gunn S and Slade D. (2020) Planning Professionalism and the Public Interest. *Planning Theory and Practice* 21: 462–464.

Schwandt TA. (2005) On Modeling Our Understanding of the Practice Fields. *Pedagogy, Culture & Society* 13: 313–332.

Scott WR. (2008) *Institutes and Organisations: Ideas and Interests.* London, UK: Sage Publications.

Searle G and Legacy C. (2021) Locating the Public Interest in Mega Infrastructure Planning: The Case of Sydney's WestConnex. *Urban Studies* 58: 826–844.

Sinnett D, Smith N and Burgess S. (2015) *Handbook on Green Infrastructure: Planning, Design and Implementation.* Cheltenham, UK: Edward Elgar Publishing.

Smith NH. (1997) *Strong Hermeneutics: Contingency and Moral Identity.* London, UK: Routledge.

Smith NH. (2002) *Charles Taylor: Meaning, Morals and Modernity.* Cambridge, UK: Polity Press.

Soja E. (1996) *Thirdspace: Journeys to Los Angeles and other Real-and-Imagined Places.* Oxford, UK: Blackwell.

Solomon D. (2003) MacIntyre and Contemporary Moral Philosophy. In: Murphy MC (ed) *Alasdair MacIntyre.* Cambridge, UK: Cambridge University Press, 114–151.

Stavrakakis Y. (2011) The Radical Act: Towards a Spatial Critique. *Planning Theory* 10: 301–324.

Stein SM and Harper TL. (2005) Rawls's 'Justice as Fairness': A Moral Basis for Contemporary Planning Theory. *Planning Theory* 4: 147–172.

Stevenson CL. (1963) *Facts and Values: Studies in Ethical Analysis.* New Haven, Connecticut: Yale University Press.

Swyngedouw E. (2009) The Antinomies of the Postpolitical City: In Search of a Democratic Politics of Environmental Production. *International Journal of Urban and Regional Research* 33: 601–620.

Tait M. (2011) Trust and the Public Interest in the Micropolitics of Planning Practice. *Journal of Planning Education and Research* 31: 157–171.

Tait M. (2016) Planning and the Public Interest: Still a Relevant Concept for Planners? *Planning Theory* 15: 335–343.

Tait M and Campbell H. (2000) The Politics of Communication between Planning Officers and Politicians: The Exercise of Power through Discourse. *Environment and Planning A* 32: 489–506.

Tanke JJ. (2011) *Jacques Rancière: An Introduction.* London, UK: Continuum.

Taylor C. (1985a) *Philosophical Papers: Volume 1, Human Agency and Language*: Cambridge University Press.

Taylor C. (1985b) *Philosophical Papers: Volume 2, Philosophy and the Human Sciences*. Cambridge, UK: Cambridge University Press.
Taylor C. (1989) *Sources of the Self: The Making of the Modern Identity*. Cambridge, MA: Harvard University Press.
Taylor C. (1995a) A Most Peculiar Institution. In: Altham JEJ and Harrison R (eds) *World, Mind, and Ethics: Essays on the Ethical Philosophy of Bernard Williams*. Cambridge, UK: Cambridge University Press.
Taylor C. (1995b) *Philosophical Arguments*. Cambridge, MA: Harvard University Press.
Taylor C. (2002a) Modern Social Imaginaries. *Public Culture* 14: 91–124.
Taylor C. (2002b) On Identity, Alienation and the Consequences of September 11th. *Acta Philosophica Fennica* 71: 165–195.
Taylor C. (2004) *Modern Social Imaginaries*. London, UK: Duke University Press.
Taylor D. (2011) Introduction: Power, Freedom and Subjectivity. In: Taylor D (ed) *Michel Foucault: Key Concepts*. London, UK: Routledge, 1–9.
Taylor N. (1994) Environmental Issues and the Public Interest. In: Thomas H (ed) *Values and Planning*. Abingdon, UK: Routledge, 87–115.
Taylor N. (1998) *Urban Planning Theory since 1945*. London, UK: SAGE Publications.
Tewdwr-Jones M. (2002) Personal Dynamics, Distinctive Frames and Communicative Planning. In: Allmendinger P and Tewdwr-Jones M (eds) *Planning Futures: New Directions for Planning Theory*. Abingdon, UK: Routledge, 65–92.
Tewdwr-Jones M. (2012) *Spatial Planning and Governance: Understanding UK Planning*. Basingstoke, UK: Palgrave Macmillan.
Thayer HS. (1981) *Meaning and Action: A Critical History of Pragmatism*. Indianapolis, IN: Hackett Publishing Company.
Thomas H. (1994) *Values and Planning*. Aldershot, UK: Ashgate Publishing.
Thomas H and Healey P. (1991) *Dilemmas of Planning Practice: Ethics, Legitimacy, and the Validation of Knowledge*. Avebury, UK: Aldershot.
Throgmorton JA. (1996) *Planning as Persuasive Storytelling: The Rhetorical Construction of Chicago's Electric Future*. Chicago, IL: University of Chicago Press.
Tubridy M and O Riain G. (2002) *Preliminary Study of the Needs Associated with a National Ecological Network*. Wexford, Ireland: Environmental Protection Agency.
UF and IEEM. (2010) *Green Infrastructure: A Quality of Life Issue*. Dublin, Ireland: Urban Forum and the Institute of Ecology and Environmental Management.
van Dijk T. (2021) What Collaborative Planning Practices Lack and the Design Cycle Can Offer: Back to the Drawing Table. *Planning Theory* 20: 6–27.
Van Puymbroeck N and Oosterlynck S. (2014) Opening Up the Post-Political Condition: Multiculturalism and the Matrix of Depoliticisation. In: Wilson J and Swyngedouw E (eds) *Post-Political and its Discontents: Spaces of Depoliticisation, Spectres of Radical Politics*. Edinburgh, Scotland, UK: Edinburgh University Press, 87–108.
Van Wymeersch E, Oosterlynck S and Vanoutrive T. (2019) The Political Ambivalences of Participatory Planning Initiatives. *Planning Theory* 18: 359–381.
Vanolo A. (2014) Smartmentality: The Smart City as Disciplinary Strategy. *Urban Studies* 51: 883–898.
Verma N. (1993) Metaphor and Analogy as Elements of a Theory of Similarity for Planning. *Journal of Planning Education and Research* 13: 13–25.
Verma N. (1996) Pragmatic Rationality and Planning Theory. *Journal of Planning Education and Research* 16: 5–14.
Verma N. (1998) *Similarities, Connections, and Systems: The Search for a New Rationality for Planning and Management*. Lanham, MD: Lexington Books.

Vigar G. (2012) Planning and Professionalism: Knowledge, Judgement and Expertise in English Planning. *Planning Theory* 11: 361–378.
Wachs M. (1985) *Ethics in Planning*. Piscataway, NJ: Transaction Publishers.
Wachs M. (1995) Foreword. In: Hendler S (ed) *Planning Ethics: A Reader in Planning Theory, Practice, and Education*. Piscataway, NJ: Transaction Publications, xiii–xv.
Wachs M. (2016) The Past, Present, and Future of Professional Ethics in Planning. In: Fainstein SS and DeFilippis J (eds) *Readings in Planning Theory*. Chichester, UK: John Wiley & Sons, 464–479.
Wagenaar H. (2011) "A Beckon to the Makings, Workings and Doings of Human Beings": The Critical Pragmatism of John Forester. *Public Administration Review* 71: 293–298.
Walzer M. (1983) *Spheres of Justice: A Defense of Pluralism and Equality*. New York: Basic Books.
Watson V. (2003) Conflicting Rationalities: Implications for Planning Theory and Ethics. *Planning Theory & Practice* 4: 395–407.
Wenger EC. (1998) *Communities of Practice: Learning, Meaning, and Identity*. Cambridge: Cambridge University Press.
Wenger EC and Snyder WM. (2000) Communities of Practice: The Organizational Frontier. *Harvard Business Review* 78: 139–145.
Wenger-Trayner E and Wenger-Trayner B. (2015) *Communities of Practice: A Brief Introduction*. Available at: https://wenger-trayner.com/wp-content/uploads/2015/04/07-Brief-introduction-to-communities-of-practice.pdf.
Wezemael JV. (2008) The Contribution of Assemblage Theory and Minor Politics for Democratic Network Governance. *Planning Theory* 7: 165–185.
Wilson J and Swyngedouw E. (2014) *Post-Political and its Discontents: Spaces of Depoliticisation, Spectres of Radical Politics*. Edinburgh, Scotland, UK: Edinburgh University Press.
Wittgenstein L. (1953) *Philosophical Investigations*. Oxford, UK: Blackwell.
Yiftachel O. (1998) Planning and Social Control: Exploring the Dark Side. *Journal of Planning Literature* 12: 395–406.
Yiftachel O. (2009) Critical Theory and 'Gray Space': Mobilization of the Colonized. *City (London, England)* 13: 246–263.
Yiftachel O, Little J, Hedgcock D, et al. (2001) *The Power of Planning: Spaces of Control and Transformation*. Dordreacht, The Netherlands: Kluwer Academic.
Young IM. (1993) Together in Difference: Transforming the Logic of Group Political Conflict. In: Squires J (ed) *Principled Positions and the Rediscovery of Value*. London, UK: Lawrence and Wishart.
Young IM. (2000) *Inclusion and Democracy*. Oxford, UK: Oxford University Press.
Young IM. (2003) The Ideal of Community and the Politics of Difference. In: Farrelly C (ed) *Contemporary Political Theory: A Reader*. London: Sage, 195–204.
Zanotto JM. (2019) Detachment in Planning Practice. *Planning Theory & Practice* 20: 37–52.
Žižek S. (1989) *The Sublime Object of Ideology*. London, UK: Verso.
Žižek S. (1997) *The Plague of Fantasies*. London, UK: Verso.

INDEX

Note: Page numbers followed by "n" refer to notes.

advocacy planning 7, 41
After Virtue (MacIntyre) 70
agon 55
agonism 56, 79
agonistic planning: *vs.* collaborative planning 75; theory 57
Alexander, E. 3–5
Allmendinger, P. 16n5, 41, 42, 50, 53–54
antagonism 55–56
Aristotle 11, 74, 79, 117
authentic dialogue 42

Banfield, E. 2–3
Beauregard, R.A. 9
Blanco, H. 33
Booher, D.E. 16n3, 42
Boyer, C. 50
branching incrementalism 29

Campbell, H. 3–5, 7, 12, 47n2, 62, 118
'character' 61, 62
Chettiparamb, A. 5–8
Chicago School 21
collaborative planning 36–44, 75, 79
Collaborative Planning (Healey) 37
common good 16n1, 19, 49, 51, 60, 61, 63; advancing 100–113; conceivable 21–47; critical approach 33; empirical research 6–8; long-term 1; planner and 87–99, 116–119; planning for 2–10, 69–86, 116–119; theoretical approach 10–12; theoretical typologies 2–6
communicative action 9, 38
communicative planning 3, 4, 40, 63
communicative practice 4
communitarianism 4, 98
'communities of practice' theory 95
complexity theory 5, 7, 8, 17, 63
conceivable common good 21–47; conceptual conundrum 45–47; deliberation and diversity 27–44; planning, as rational scientific activity 21–27
conflict management 57
conflictual consensus 61
consensus-seeking planning processes 60
contractarianism 87
Cosmopolis 40–41
counter-conducts 52, 53
critically uncommon good 48–65; critical planning theory 48–49; forgetting 61–64; forsaking 61–64; Foucauldian critique 49–53; post-political critique 53–61
critical planning theory 48–49
critical pragmatism 30–36, 40
cultural work 52, 53

Deleuze, G. 10, 19n1, 63
deliberation: and diversity 27–44; inclusive 13, 18; moral 75, 85n2; reflective 15

deontology 16n1
desires, first- /second-order 90
Dewey, J. 7, 8, 29–31, 33
dialogical concept of public interest 3, 4
dialogical proceduralism 5
discursive ethics 46
disjointed incrementalism 35
'distribution of the sensible' 58, 59
divergentism 5
diversity 27–44, 63

emotivism 85n2
Enlightenment project 70, 73, 88
equality 33, 34, 56–59
ethics: definition of 53; discursive 46; inclusionary 42, 46, 47n3; of the self 53

Fainstein, S. 43–44, 46, 47n6
Faludi, A. 26
first-order desires 90
Flyvbjerg, B. 51
Forester, J. 9, 31–33, 35, 37, 39, 43, 47n5, 50
formalism 89, 96
Foucault, M. 48–53, 79; critique of common good 49–53; planning theories 63
Frankfurt, H. 90
Friedmann, J. 29, 35

genealogical method 72
Giddens, A. 37–38
GI *see* green infrastructure (GI)
governmentality 52, 58
green infrastructure (GI) 15, 101, 102, 104–105, 107–109
Gunder, M 5, 10, 50, 62, 65n2

Habermas, J. 21, 31–32, 36, 40–43, 45, 47n4, 56, 60, 69, 89
Harper, T.L. 16n3, 34, 35
Harvey, D. 47n6, 57
Healey, P. 9, 18, 26–27, 31, 36–39, 95
hegemony 58
Hillier, J. 10, 26–27, 62, 65n2–3
Hoch, C. 27, 30, 36, 37
Howe, E. 3–5
Huxley, M. 50, 52, 53

ideal speech 31–32, 42
identity 61–62, 84, 118
ignorance: persistence of 69–70; veil of 34
Inch, A. 9, 52
inclusionary ethic 42, 46, 47n3

incrementalism: branching 29; disjointed 35
individualism 3, 4, 10, 92
Innes, J.E. 16n3, 29, 42
institutionalism 17, 95

Jacobs, J. 22, 77, 78, 86n3
James, W. 27, 28, 33, 34
'Just City' model 43, 44, 46
justice, substantive theory of 44; *see also* social justice

Kant, I. 10, 16n1, 71, 87, 89
knowledge: of community 34; production 25, 28, 36; of truth 34
Kühn, M. 56, 57

Laclau, E. 54
legitimacy 22, 32, 114; of evaluations 6; moral 97; of planning 30, 78
Lindblom, C.E. 21, 29, 35
Lutz, C.S. 81, 85n1

MacIntyre, A. 10–12, 15, 67, 70–77, 79–85, 85n1, 85–86n2, 88, 89, 95, 115–117
Maidment, C. 7, 8
Marshall, R. 3–5, 7, 47n2
Marxism 43, 70
Marx, K. 10
McLoughlin, B. 22
Mill, J.S. 3, 10, 16n1, 69, 71
mixed-scanning 29
morality/moral: fiction 87–88; legitimacy 97; pluralism 15; rationality 71; realism 89, 90–95, 115
Moroni, S. 5, 16n3–4
Mouffe, C. 54–58, 60, 61

neoliberalism 7, 47n3, 52, 58, 60
neo-pragmatism 34
Nietzsche, F. 10, 48, 50, 51, 55, 72–74, 76, 88, 89
nomocracy 5

partial impartiality 46
persistence of ignorance 69–70
'phronetic' approach 51
Pierce, C. 28, 33
planning: advocacy 7, 41; agonistic 57, 75; big 21; collaborative 36–44, 75, 79; for common good 114–119; communicative 3, 4, 40, 63; consensus-seeking processes 60; critical theory 48–49; as disciplinary technology 50;

land-use 100; legitimacy of 30, 78; moral landscape, contouring 95–98; as practice 79–84; public interest 2–10; rationality 51; rational process 64; as rational scientific activity 21–27; as scientific activity 40; subject 8–10; theory 2, 3, 5, 12, 17–19, 53, 63, 64, 65n5, 69, 82, 87, 115; as tradition 74–79, 87, 89, 111–112; transactive 29, 35; transforming practices in 102–111
police 58, 59, 65n4
police order 58–61
political, the 55
political theory 5, 7, 31
politics 56, 59; democratic 55; of moral deliberation 75–76
post-political theory 61, 63
power 30, 38, 49–51, 79; accumulation 26; asymmetries of 1, 31, 36, 43, 46, 49, 50; computational 22; dimensions 44, 50, 52; distortions of 32, 33, 43, 47n5; hidden exercise of 51; hierarchies 43–44; issues of 13; will to 10, 88
practice: characteristics of 81; definition of 80; self-referential 82
pragmatism 27–30; critical 30–36; neo- 34
proceduralism 115; dialogical 5; rational 36
public interest 1–2, 16n1, 41, 51; dialogical concept of 3, 4; empirical research 6–8; intersubjective perspective of 4; planning 2–10; theoretical approach 10–12; theoretical typologies 2–6

quest 85, 102, 116, 117, 118

Rancière, J. 54, 57–61, 65n4–5
rationality: moral 71; planning 51; substantive 85; technical-instrumental 36
rational process: planning 64; theory 24–27, 39, 49; thinking 62
Rawls, J. 5, 10, 34, 47n2, 57, 69, 89

realism/reality 77, 87–88
relational ontological approach 38
relativism 77
repressive hypothesis 50
Ricoeur, P. 13, 61–62, 64n1
right, and good 88–89

Sager, T. 36, 41, 47n3
Sandercock, L. 9, 40, 41
Schmitt, C. 55
Schön, D. 30
second-order desires 90
selfhood 61, 65n5
situated ethical judgement 115, 118, 119
social constructionism 28, 31
social-constructivist approach 38, 49
social imaginary 97–98
social justice 5, 41, 44, 93
Stein, S.M. 16n3, 34, 35
structuration theory 38, 95
subjectification 52, 59–61
substantive-focused theories 16n3
systems theory 22–24, 45, 79

Taylor, C. 11, 15, 67, 88–93, 95–98, 98n2, 115
transactive planning 29, 35

universalism 77, 115
utilitarianism 3, 4, 16n1, 69, 71, 96

Verma, N. 33–34

webs of interlocution 11
Wenger, E.C. 95

Yiftachel, O. 50
Young, I.M. 5, 12

Zanotto, J.M. 9
Žižek, S. 9, 64–65n2